Adobe Illustrator for Creative Professionals

Develop skills in vector graphic illustration and build a strong design portfolio with Illustrator 2022

Clint Balsar

BIRMINGHAM—MUMBAI

Adobe Illustrator for Creative Professionals

Copyright © 2022 Packt Publishing

Group Product Manager: Rohit Rajkumar
Publishing Product Manager: Rohit Rajkumar
Senior Editor: Hayden Edwards
Content Development Editor: Rashi Dubey
Technical Editor: Joseph Aloocaran
Copy Editor: Safis Editing
Project Coordinator: Ajesh Devavaram
Proofreader: Safis Editing
Indexer: Hemangini Bari
Production Designer: Ponraj Dhandapani
Marketing Coordinator: Elizabeth Varghese

First published: June 2022

Production reference: 1300622

Published by Packt Publishing Ltd.
Livery Place
35 Livery Street
Birmingham
B3 2PB, UK.

ISBN 978-1-80056-925-6

www.packt.com

To my students; past, present, and future. It is for you that I strive to accomplish great things and then inspire you through my example. I hope you find this book helpful and encouraging. This project was my latest attempt to remind you not to be afraid of challenges; rather to rise to the occasion and complete things that may first seem difficult and risky.
A strong determination will see you through the task and allow you to learn so much along the way. Remember that it is not only alright to make mistakes, but that they are necessary for growth. We learn far more from the actions we don't want to repeat than from those we do.

– Clint Balsar

Contributors

About the author

Clint Balsar is an experienced creative with a demonstrated history of working in the primary/secondary education industry. Clint is skilled in digital illustration and graphic design, as well as educational technology, instructional design, and Adobe and Apple products. He possesses varied certifications in Adobe products and training skills.

I want to thank all my friends and family for being so supportive during the time of authoring this book, especially my incredible wife, Ticia. Her love and support made all the difference as I taught my digital arts students each day, attended numerous meetings after school, and then wrote this book's content during most of the available evenings and weekends.

About the reviewers

Craig Daalmeijer-Power has worked as a multidisciplinary creative for more than 30 years. He has worked on projects for local communities and clients and worked in the intelligence community in support of the Australian Defence Force. He received the award of Adobe Education Leader of the Year in 2021 for his commitment to excellence in creative education. His work within the Adobe education community is highlighted by regular international presentations, Behance streams, and Adobe Creative Career mentoring. He is currently employed by TAFE NSW, as a full-time head teacher, where he manages courses in photography, screen and media, and games development and focuses on innovation and creativity whilst taking pride in being a lifelong learner.

Thank you is the simplest form of credit and to me, it is simply being open to being thankful and surrounded by inspiration in everything we involve ourselves with. This inspiration for me is the supportive communities within Adobe Creative educators, Adobe Education leaders, and my family, both at home and work. I am grateful to all the members of each of these communities, my extended family, and how they inspire me to always achieve my best.

Najihah Najlaa is an **Adobe Certified Expert** (ACE) and professional certification holder and an instructor. In addition to her profession as an instructor, Najihah is also the founder of her own video production company, "A Lifetime Project," and is a sought-after freelance photographer, videographer, and designer. Known as a person who moves with passion, Najihah has dived deep into the Malaysian creative industry fully equipped with her in-depth knowledge and Adobe certifications. She has been conducting and facilitating various workshops and Adobe training nationwide for the last six years.

Since 2020, she has been actively producing creative project on "Behance livestream" and is a Behance Featured Stream Artist.In 2021, she was invited to speak at Adobe Max 2021.

She is currently managing her own company and conducts creative trainings for the corporate and private sectors to help creative people polish their skills with comprehensive, high quality training covering Adobe media products, photography and videography.

I am overwhelmed in all humbleness to acknowledge my debt to all those who have helped, inspired, and given me the golden opportunity to this wonderful project, which also helped me in doing a lot of research in which I learned about so many new things. I am really thankful to them. Finally, to my caring, loving, and supportive cats, Kedo and LJ: my deepest gratitude for accompanying me when the times got rough. It was much appreciated and duly noted.

Table of Contents

2
Prepping for Illustrator

Part 2 – Advanced Illustrator Methods

3
Developing and Organizing Objects

4
Drawing with the Pencil, Paintbrush, Pen, and Shape Tools

5

Editing and Transforming Objects

6

Advanced Attribute Design

7

Powerful Typography Options in Adobe Illustrator

Part 3 – Real-World Applications

8

Preparing Artwork for Presentation

9

Utilizing Multiple Artboards

10

Advanced Layer Development and Organization

11

Extending Illustrator Through Third-Party Tools and the iPad

12

Illustrator Mastery – Advanced Techniques and Shortcuts

Index

Other Books You May Enjoy

Preface

Adobe Illustrator is a vector-based art tool for visual creatives. It is an industry-standard tool that allows a design to go from concept to completion, including the process of peer collaboration and client feedback.

Complete with step-by-step explanations of essential concepts and practical examples, you'll begin to build confidence as you master the methods of successful illustrators in the industry by exploring the crucial tools and techniques of Adobe Illustrator. You'll learn how to create objects using different tools and methods while assigning varied attributes and appearances. Throughout the book, you'll strengthen your skills in developing structures for maintaining organization as your illustration grows.

By the end of this Adobe Illustrator book, you'll have gained the confidence you need to not only create content in the desired format and for the right audience but also build eye-catching vector art based on solid design principles.

Who this book is for

This book is for creative illustrators with basic to intermediate-level experience with vector-based software who want to take their existing skills to the next level. Prior knowledge of vector-based illustration concepts will help you to get the most out of this Adobe Illustrator software book and produce impressive results.

What this book covers

Chapter 1, Building a Foundation beyond the Basics, explores how creators of all levels can still benefit from a review of the ins and outs of the advanced features and options available within Adobe Illustrator. This chapter can quickly check your confidence level and raise it even higher, as you prepare to delve deeper into mastering this software.

Chapter 2, Prepping for Illustrator, discusses how initializing content in Adobe Illustrator can come in many forms. A quick review will allow you to feel empowered to develop original content that is kick-started from sketches and outside resources.

Chapter 3, Developing and Organizing Objects, explains how although organization is not the most exciting thing to talk about, it is arguably one of the most important. Artists that harness their creations in clear, concise layers and groups will spend much less time fixing objects later.

Chapter 4, Drawing with the Pencil, Paintbrush, Pen, and Shape Tools, presents several methods for content creation. You will discover that although you have learned to use Adobe Illustrator in one or more ways, more options are still available. Creativity will expand after experiencing the variety of methods this software allows for.

Chapter 5, Editing and Transforming Objects, explains how creating more advanced objects is only part of the battle in mastering Adobe Illustrator. A successful creator must also have advanced skills in manipulating the object through transformations and editing methods. Effects and their options can also be a real time-saver in advancing the look of your work.

Chapter 6, Advanced Attribute Design, explains how even though you will have learned pretty early that each path can obtain two key attributes (fill and stroke), there is far more to learn about the optional appearance attributes you can assign to any object.

Chapter 7, Powerful Typography Options in Adobe Illustrator, intends to offer a higher level of detail on using and manipulating typography. We will look at the benefits of Adobe Fonts and what quality fonts offer the designer. Additional information on typography skills offered by Adobe Illustrator will also be explored in this chapter.

Chapter 8, Preparing Artwork for Presentation, discusses how there are many methods to develop your vector artwork in preparation to present it, but as it is not resolution-dependent, it can also be adjusted for multiple needs. We will explore several options for output while putting a focus on "raising the bar" as to its professional appearance.

Chapter 9, Utilizing Multiple Artboards, explores how the power of customized artboards and multiple artboards in a single file allow designers to collect a project in one central location.

Chapter 10, Advanced Layer Development and Organization, covers the relationship between organized layers and multiple artboards and shows strategies for maintaining control of all assets within the project. Using layers, sublayers, and their colors to keep track of advanced projects, creators maintain greater control of their work and become more efficient over time.

Chapter 11, Extending Illustrator through Third-Party Tools and the iPad, discusses how even though Adobe Illustrator is already an amazing tool for developing original work, there are a few places to find both free and paid for enhancements. The Adobe Exchange is one such place that we will explore. In addition to highlighting a few selected plugins with a specific task or enhancement, we will also be reviewing the more complex plugin sets available from Astute Graphics.

Chapter 12, Illustrator Mastery – Advanced Techniques and Shortcuts, explains how to apply your new-found knowledge and skills through a series of challenges with supplied sample files. Discussion on best methods for utilizing Adobe Illustrator and the user's Creative Cloud will focus on continued organization and potential collaboration.

To get the most out of this book

You will need a version of Adobe Illustrator installed on your computer – the latest version, if possible. You will also benefit from having Astute Graphics plugins and the Maxon Cineware for Illustrator plugin (although not required).

Software/hardware covered in the book	Operating system requirements
Adobe Illustrator 2022 (Version 26 or higher)	Windows and macOS
Astute Graphics plugins	
Maxon Cineware for Illustrator plugin	

There will also be opportunities to increase your knowledge of Illustrator for iPad, so an Apple iPad will allow you to get even more out of this book.

Download the color images

We also provide a PDF file that has color images of the screenshots and diagrams used in this book. You can download it here: `https://packt.link/2EucI`.

Conventions used

There are a number of text conventions used throughout this book.

`Code in text`: Indicates code words in text, database table names, folder names, filenames, file extensions, pathnames, dummy URLs, user input, and Twitter handles. Here is an example: "The **Actions** panel allows for a series of prerecorded steps, while the **Variables** panel allows the use of data files (`.cvs` or `.xml`) to replace the content with a document."

Bold: Indicates a new term, an important word, or words that you see onscreen. For instance, words in menus or dialog boxes appear in **bold**. Here is an example: "You can also adjust the **Fidelity** slider to change the path from being more accurate to being smoother."

> **Tips or Important Notes**
> Appear like this.

Get in touch

Feedback from our readers is always welcome.

General feedback: If you have questions about any aspect of this book, email us at customercare@packtpub.com and mention the book title in the subject of your message.

Errata: Although we have taken every care to ensure the accuracy of our content, mistakes do happen. If you have found a mistake in this book, we would be grateful if you would report this to us. Please visit www.packtpub.com/support/errata and fill in the form.

Piracy: If you come across any illegal copies of our works in any form on the internet, we would be grateful if you would provide us with the location address or website name. Please contact us at copyright@packt.com with a link to the material.

If you are interested in becoming an author: If there is a topic that you have expertise in and you are interested in either writing or contributing to a book, please visit authors.packtpub.com.

Share Your Thoughts

Once you've read *Adobe Illustrator for Creative Professionals*, we'd love to hear your thoughts! Scan the QR code below to go straight to the Amazon review page for this book and share your feedback.

https://packt.link/r/1-800-56925-4

Your review is important to us and the tech community and will help us make sure we're delivering excellent quality content.

Part 1 – Reviewing the Necessary Knowledge

This part will ensure that you have the necessary understanding to help you feel more prepared for the remainder of the book and will help you to acquire the necessary knowledge and/or skill before advancing.

This part comprises the following chapters:

- *Chapter 1, Building a Foundation beyond the Basics*
- *Chapter 2, Prepping for Illustrator*

1
Building a Foundation beyond the Basics

Adobe Illustrator has been the industry leader in vector drawing software for decades and continues to be a very popular choice for graphics professionals. As you begin your journey into this book, it is very likely that you have been working with Illustrator for some time now but want to get more from the software and raise your skills to a higher level. It might also be the case that you haven't worked with Illustrator for very long and hope to hone your early skills.

Whatever the reason, creators of all levels can still benefit from a review of the ins and outs of the advanced features and options available within Adobe Illustrator. This chapter can quickly check your confidence level and raise it even higher as you prepare to delve deeper into mastering this software. Although this is not going to be a thorough list of all the foundational knowledge you would need in order to master Illustrator, it is intended to get you on stable footing as we move forward in this journey of harnessing its power.

After reading this chapter, you'll understand how to organize and save customized workspaces. You will also know how to utilize the Properties panel and Control panel to select attributes and become more efficient with the use of panels and shortcuts for often-used tools and tasks. You will be able to stay organized by using a grid, guides, and/or smart guides, and finally, keep up to date with the latest enhancements offered for Illustrator.

To accomplish this, the chapter will be divided into the following main topics:

- Creating a file.
- Workspaces and preferences.
- Tools and the Control panel.
- Additional panels and shortcuts.
- The grid and guides.
- What's new in Adobe Illustrator 26.3.1?

Technical requirements

To complete this chapter, you will need the following:

- Adobe Illustrator 2022 (version 26.0 or above).
- High-quality internet access may be required for some situations.

Creating a file

In this first chapter, we will be reviewing some of the foundational bedrock of Adobe Illustrator that has allowed it to maintain its reputation as a powerhouse vector-based arts software for so many years. Control and customization are two key components of Adobe Illustrator that make it flexible for any designer's needs.

Once you have initiated Illustrator, you will be greeted with a welcome screen as follows:

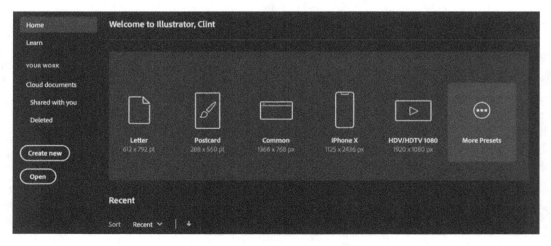

Figure 1.1 – Home screen options

This welcome screen will allow you to create new files, open previous ones (**Recent** is displayed prominently, while links are also available for **Cloud documents** and **Shared with you**), and even connect to Adobe's vast learning library. The latter offers both hands-on tutorials (in-app) and tutorials on the web. In addition to these options, you will also find a **What's new** button at the bottom left that Adobe updates with each new release of its software.

If you are starting a new file, the **New Document** dialog box allows you to choose from **Presets** based on your intended output (print or screen), **Templates** from Adobe Stock, or creating your own custom files, as follows:

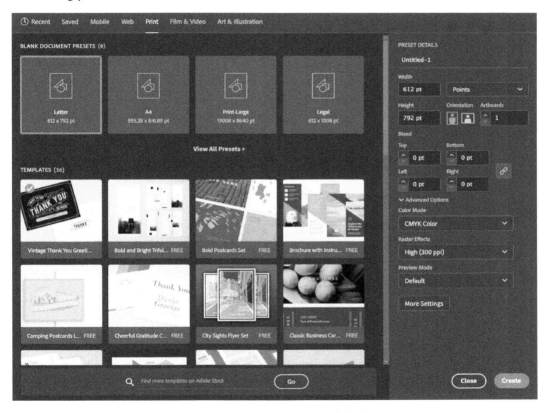

Figure 1.2 – The New Document dialog box

The presets are divided into the media categories of **Mobile**, **Web**, **Print**, **Film & Video**, and **Art & Illustration**. In addition, Illustrator also offers a category for **Recent** presets and **Saved** templates that you have acquired from Adobe Stock.

Workspaces and preferences

How you lay out your **workspaces** and develop that creative workflow will assist you in being more focused and efficient. Workspaces allow you to have the most common tools for the intended task while reducing unnecessary clutter.

Once you have greater control of your workflow management, you can then continue to make refined adjustments through the software's **Preferences** panel. This will give you the professional level of control that you will inevitably seek as you become more and more confident with the advanced options in Illustrator.

Workspaces

Illustrator currently offers the user nine different workspaces (plus **Touch** if your screen allows for it). These are merely a redistribution and/or visibility of the panels that would best be suited for the intended task. This allows the **graphical user interface (GUI)** to be reconfigured for the job you are performing or what feels most comfortable to you as the user. It also allows for the creation of custom spaces, if you would like to set up a unique environment that is best suited for you.

Workspaces can be selected in the Application bar located at the top right of the software's interface, as shown here:

Figure 1.3 – Workspace menu

As you can see in *Figure 1.3*, the **Workspace** menu has quite a few choices available. **Essentials** and **Essentials Classic** are great general-use layouts, but Illustrator also offers several workspaces for more specific tasks. Let's have a quick review of each one and what it offers:

- **Automation**: This offers up the panels you would like when doing repetitive work. The **Actions** panel allows for a series of prerecorded steps, while the **Variables** panel allows the use of data files (`.cvs` or `.xml`) to replace the content with a document.

- **Essentials**: This is the default view Illustrator opens with and simplifies the GUI to just three key panels: **Properties**, **Layers**, and **Libraries**.

- **Essentials Classic**: The previous default view that Illustrator used to open with and offers a good variety of the most common panels.

- **Layout**: This workspace is for those times when accuracy is crucial. The **Transform** and **Align** panels appear at the top right of the interface. It also allows access to multiple panels for object attributes, such as **Fills**, **Strokes**, and **Gradients**, as well as several panels for modifying the typography within the document.

- **Painting**: This offers the panels you would likely want for a painting project, including **Brushes**, **Color**, and **Navigator**.

- **Printing and Proofing**: If you are taking your document to the press, you may want to utilize this workspace. The key panel in this group is **Separations Preview**. It allows you to see your overprint preview by dividing your CMYK color space.

- **Tracing**: This allows you to image trace a raster image and adjust it for improved quality.

- **Typography**: Again, as the name would imply, you will find all the panels needed to make advanced adjustments to text.

- **Web**: This workspace has several familiar panels while adding the CSS **Properties** panel for those working with web content.

As well as these workspaces, customized workspaces can be saved and will appear at the top of the menu too.

Custom workspaces

To save a custom workspace, follow these steps:

1. After creating a new document, open all the desired panels and place them where you would like them to be.

2. Go up to the **Switch Workspace** button located at the top right of the Application bar and select **New Workspace…**.

3. Enter a custom name for your new workspace.

4. You should now find this and any additional custom workspaces at the top of the list when returning to the **Switch Workspace** button or when using the **Window** > **Workspace** menu.

For those unfamiliar with arranging their own workspace, let's review the concept of the **dock**. Illustrator panels and panel groups can be collected in long vertical arrangements known as **docks**. To dock a panel, you will select it from the **Windows** menu at the top of the screen. Once it is open, it is considered a floating panel. To add it to an existing dock, you will need to grab it from the tab that includes the panel's name. This will be located at the top of the panel. You can then drag it over either a dock or a panel group to place it there.

Figure 1.4 shows what it would look like if you were about to add it to an existing panel group (left image), add it to an existing dock but outside a panel group (center image), and the result of adding it to an existing dock but outside a panel group (right image):

Figure 1.4 – Docking a panel

Custom workspaces are a great way to make Illustrator work for your intended purposes. I like to create custom workspaces for my multiple screen set-ups and third-party software. They are easy to create and remove, so don't be afraid to treat them as temporary aids to your workflow.

Custom preferences

Using the *Ctrl/Command + K* shortcut will bring up your preferences in several pieces of Adobe software (including Illustrator) and is an excellent place to start customizing Illustrator for your specific usage. From here, you can decide how many levels of **Undo** will get recorded, attach to a scratch disk, and set the time intervals between automatic recovery saves.

An excellent example of this is the **Scale Strokes & Effects** command located under the **General** section of this menu. I usually check this item so that the scaling of an object stays visually consistent at any size. I will be referring to this panel quite often through the course of this book, but for now, it is most important to know that it is available and allows for greater flexibility and customization for the user.

Contextual menus

An excellent option for quick access to commonly used tasks is the use of **contextual menus**. By definition, a contextual menu, or shortcut menu, gives you access to frequently-used commands related to the current context. It is generally brought up through a shortcut key or mouse button In Illustrator, after making a selection, you can right-click to see options for the selection:

Figure 1.5 – Contextual menu to an existing dock, but outside a panel group

In the case of the contextual menu shown in *Figure 1.5*, it highlights the options available to the path with the dashed white stroke. You will notice that it is the selected object due to the visibility of its blue **bounding box**. After summoning the contextual menu, you can use the **Transform** command to quickly make adjustments, such as adjusting the scale, and then either change the original selection or make a copy by using the **Copy** button.

Object attributes

At its simplest, every object can be assigned two attributes: **Fill** and **Stroke**. Of course, today, a variety of additional options can also be used to enhance any given object. In fact, an object can have multiple fills and strokes. You can adjust stroke width as well as apply variable stroke widths, brush definitions, and opacities.

Most of the adjustments to a path's attributes can be adjusted in the **Control** panel, located directly below the Application Bar (except in the default **Essentials** workspace). To assign more advanced attributes, such as adding multiple strokes to a path, you will need to make your way to the **Appearance** panel (which can be accessed through the **Properties** panel):

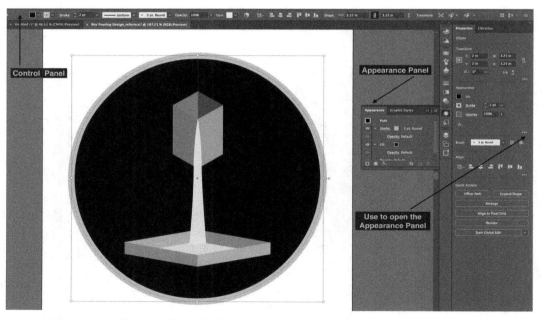

Figure 1.6 – The Control panel offers a lot more than just the Fill and Stroke attributes

As you may already be familiar with, the **Fill** attribute can be anything that can be a **swatch** in your swatches library. It can be a solid color, a gradient, or a pattern (**None** is also a choice). The same three choices of color, gradient, or pattern (plus **None**) are also available for the **Stroke** attribute.

> **Important Note**
> Setting up Adobe Illustrator is like finding the perfect recipe. You will continue to adjust it over time, and soon find that what was perfect for you at one time is no longer the right "flavor" and must be again adjusted for the specific situation at hand.

Understanding the multitude of ways to customize and utilize the GUI of Adobe Illustrator will help you solidify your success as you work to master your control of software. In extending the earlier analogy of the perfect recipe, consider that as chefs gain greater success, they inevitably also get more accustomed to their cooking tools and even customize them and their uses. Just as chefs hone their skills and the use of their tools over time, you too will find more benefits for the **Workspaces** and **Preferences** options over time and use them as your "secret sauce" in creating with Illustrator.

But of course, to successfully create in Illustrator, you must repeatedly use the main tools in Illustrator, while getting to know their specific capabilities (and limitations) over time.

Tools and the Control panel

I won't be going over all the available tools in this introductory chapter but will be reviewing some of those that are vital to successfully using Adobe Illustrator. In addition, I will share the Control panel for making choices and discuss the panel that Adobe has chosen as its replacement when in the Essentials workspace.

The Bezier curve

The development of Illustrator began from the earlier success of the PostScript language developed by the founders of Adobe: John Warnock and Chuck Geschke. The original was released in 1987 (which happens to be the same year I graduated from high school) for Apple Macintosh. The first official Windows offering occurred two years later with version 2.0 in 1989. Just like its predecessor, Illustrator was based on the PostScript language, which creates vector paths defined by mathematical formulas. The smooth curves were a welcome improvement in desktop publishing from prior pixel-based solutions.

Figure 1.7 – Original Illustrator splash screen

So, the curve is fundamentally what makes Adobe Illustrator unique and essential. It is smooth at any resolution. In fact, it is generally stated that vector-based files are "resolution independent." It could be scaled to work just as well if it were for a business card or the side of a 10-story building.

The next couple of pages of tools are all based on the Bezier curve. If you are somewhat unfamiliar with all the tools and their shortcuts, you may use the following as a reference guide:

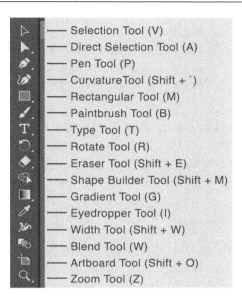

Figure 1.8 – Tools and their shortcuts in the Essentials workspace

Additional tools will be located under each of the tools that show a small triangle in the lower-right corner, and we will review all of them in later chapters.

Pen Tool

One of the most important tools to get skilled with in Adobe Illustrator is the **Pen** tool. It develops paths with either straight or curved segments between **anchor points**. It isn't always the most intuitive tool in the collection from Illustrator, but it is one of the most flexible.

Figure 1.9 – Path and anchors

Using the **Direct Selection** tool, you can edit any anchor or grouping of anchors. In addition to moving the anchor's position, you can also adjust its angle and size if it is a curvilinear segment of a path. Using the **Anchor Point** tool, you can also convert a corner anchor to a curve anchor and vice versa. To be more efficient when drawing with the **Pen** tool, several techniques can be used that will allow you to draw and edit simultaneously, and we will be going over these, and more, in the chapters to follow.

Type

Type has always been a strength of Adobe Illustrator since that is exactly how Adobe got started (think PostScript). For many years, Illustrator only had Adobe PostScript Type 1 fonts, but as the TrueType format (designed by Apple) gained in popularity, Adobe decided to allow support for both in version 7 and beyond.

This allowed content to be created for additional media, as Type 1 contains PostScript information for printing, while TrueType is a screen-appropriate font and not intended for print. Meanwhile, with the introduction of OpenType, which offers information for both print and screen, creators have found it to be the preferred method due to its flexibility.

Type tool is a collection of tools for text; besides the **Type** tool, there is also the **Type on a Path** tool and the **Vertical Type** tool.

Of course, you really can't mention the **Type** tool and not talk about the **Character panel**, which you can see here:

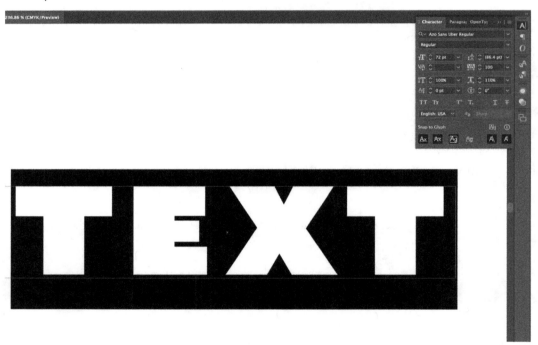

Figure 1.10 – Text editing with the Character panel

The Character panel can be made available by either selecting the **Typography** workspace or using the *Ctrl/Command + T* keyboard shortcut. Here, you can choose the font with a **What You See Is What You Get (WYSIWYG)** view as well as choose the typography **Tracking**, **Kerning**, and **Leading** elements (all elements of spacing).

Pencil tool

As a relative of the **Pen** tool, the **Pencil** tool can still be used to create vector paths but allows them to be drawn in a more natural and intuitive way. The best thing about it is that it can be edited in the exact same manner as the **Pen** tool. Both tools create paths and anchors in their results. It is only their method of creating those paths and anchors that vary.

Shape tools (Rectangle, Ellipse, Polygon, and Star)

Just like the **Pen/Pencil tools** (and similar), the shape tools make use of the Bezier curve to create their paths. Each shape can then accept the common attributes of **Fill** and **Stroke**, as well as the more advanced options and effects.

Shape Builder tool

This tool has grown to be one of my favorite tools in the entire set. After selecting a collection of overlapping shapes, you can quickly combine them by drawing a line across their intersecting paths. In addition, you can also subtract their intersecting shapes by holding down the *Alt/Option* key while clicking or drawing a line across them.

If you have been using Adobe Illustrator for many years now, you may already know the connection between this tool and the older tool known as **Pathfinder**. We will also be taking a closer look at Pathfinder and how the combination of these two tools allows for the most options.

Control panel

In the **Control panel**, you have a multitude of available visuals and choices at your disposal. To investigate just what will be shown in the panel when appropriate, right-click on the contextual menu to the far right of the panel to display the entire list of options:

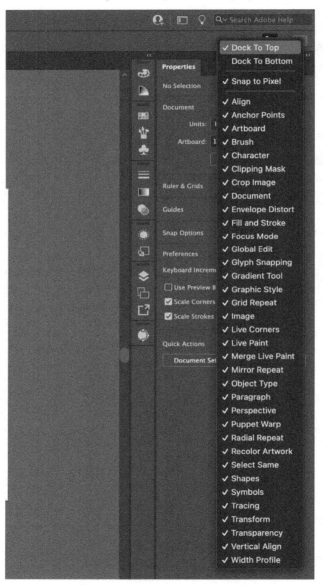

Figure 1.11 – Control panel view options

If any options are unchecked, they will not be displayed inside the Control panel, even when it is appropriate to do so.

As you can see in *Figure 1.11*, the Control panel allows for a lot of content in one small space on your screen. In addition, you can even choose to dock it to the top or bottom of the GUI.

> **Important Note**
>
> Adobe has elected to remove the Control panel from Essentials, which is their default workspace view. If you want to leverage the benefits of this panel, you can choose any of the other workspaces.

If you have been using Illustrator for quite a while you will understand why I wanted to discuss the Control panel and its usage before moving on to the newer Essentials workspace and the transition to using other panels and methods for finding information and options. If you have only been using Illustrator for a short time, then you may have been keeping within the comfort zone of the Essentials workspace, and this discussion will be helping you to venture into the additional options. Either way, it's important to acknowledge these foundational options before moving forward.

Additional panels and shortcuts

In moving forward, let's address those items that you should all have a common knowledge of. In this section, we will review the key panels for assigning attributes and editing them at any time. We will also look at a few powerful methods to keep designs organized as well as a few keyboard shortcuts you will want to know as you become more efficient while using the software.

Appearance panel

Once an object is selected, you can review and adjust assigned attributes and effects on the object (or similar objects). It will show different information based on what has been selected. If the selected object is a path, then the **Fill** and **Stroke** attributes will be displayed. It will also display **Opacity** and Blending Mode attributes, as well as effects applied to the path. This can also be where you choose to add any or all the aforementioned attributes.

When looking for the Appearance panel, look for an icon that appears to have a circle with a solid white fill and a white dashed line stroke around it, as in *Figure 1.12*:

Figure 1.12 – Appearance panel showing the properties of the selected object

You may also summon the panel by going to the menu and choosing **Window** > **Appearance**. Note that the information within this panel will change each time you select a different object, multiple objects (selecting while holding the *Shift* key), or groups.

Properties panel

The **Properties panel** is a bit of a one-stop shop for viewing attributes of a selected object and quickly adding to or modifying it. This Swiss Army knife of a panel allows for **Transform** adjustments, **Appearance** adjustments, **Alignments**, and additional **Quick Actions**, such as **Arrange** and **Recolor**. If there is currently nothing selected, the panel allows you to set up several document and user interface options, such as **Ruler & Grids**, **Snap Options**, and **Document** and **Artboard** settings:

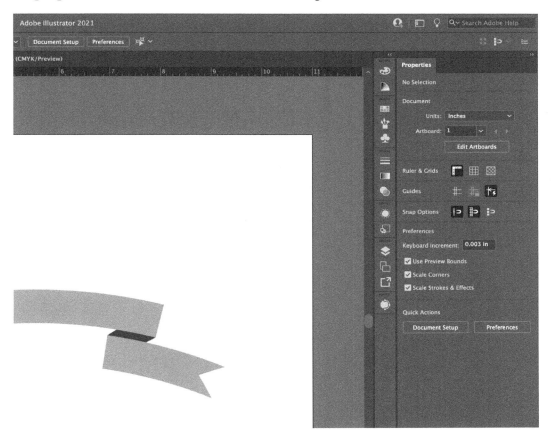

Figure 1.13 – Document properties

While using the **Selection** tool, you will find the document's properties in the **Properties** panel when nothing is selected. From this panel, you can change the unit of measurement, add artboards, and many more options. Located at the bottom of this panel will always be **Quick Actions**. This area showcases several shortcut buttons that may accelerate your efforts. In the scenario shown in *Figure 1.13*, the two buttons available to you will be **Document Setup** and **Preferences**. Once an item or items becomes selected, the panel changes to display the properties available for that specific situation:

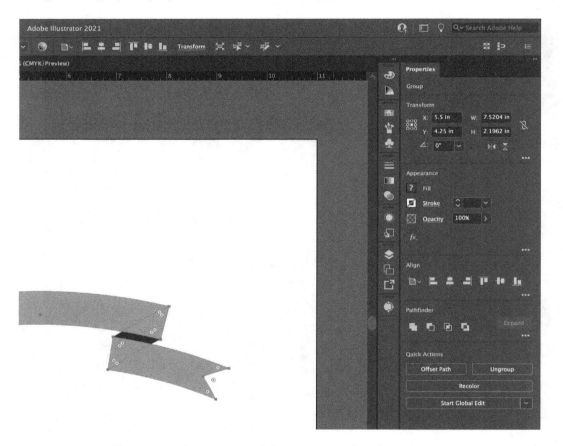

Figure 1.14 – Properties panel showing options based on selection

Several manipulations can be accomplished using this panel, including access to the aforementioned **Appearance** panel. Many of the options are available there, while any additional options can be accessed using the **More Options** button located in the lower-right corner of each section.

You should now feel comfortable with setting up and adjusting your workflow for a variety of intended tasks, while continually becoming more efficient with your process of creation. You may find that your current situation still requires a more precise layout, though. In such cases, I would highly recommend the use of the following compositional aides.

The grid and guides

One of the key methods for keeping your designs to a high professional standard is precision. By using the following layout methods in conjunction with a variety of alignment techniques (we will be going over them in future chapters), you will be able to construct with confidence.

Grid

You can quickly view (and hide) your **grid** layout by using the *Ctrl/Command + '* keyboard shortcut. In addition, if you would like to customize the grid, you can summon the **Preferences** panel with the keyboard shortcut of *Ctrl/Command + K*. One benefit to turning on the grid's visibility is the ability to snap to it. You can activate **Snap to Grid** using the keyboard shortcut of *Shift + Ctrl/Command + '* as shown here:

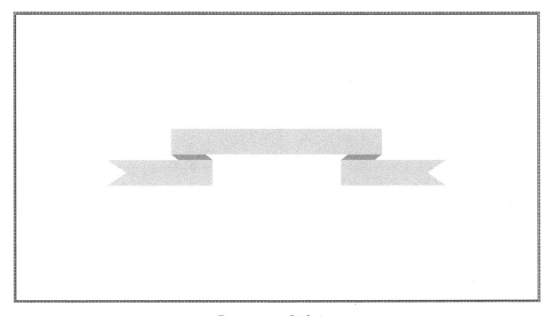

Figure 1.15 – Grid view

Be sure to use the Properties panel to customize the grid to your liking. From here, you can choose **Snap to Grid** or go into the preferences by using the **Quick Actions** button to change the dimensions, style, and color of the grid.

Rulers

After turning on the **Rulers** view (*Ctrl/Command + R*), you can further utilize the tool by adjusting the unit of measurement. Just right-click on either ruler to see a contextual menu for all the available units of measurement, and then select your preferred units:

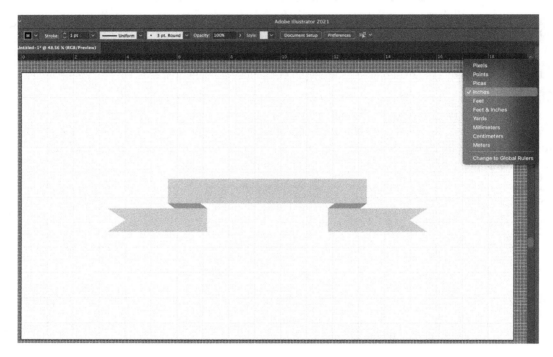

Figure 1.16 – Rulers view

I suggest changing your rulers to the unit of measurement your intended output will use. For example, if you are intending the graphic for the web or screen, then using pixels is likely the best choice. If the end result is to go to print, then the units should relate to that.

Guides

To add **guides**, just locate your cursor inside the vertical ruler to click and then drag a vertical guide and, similarly, locate your cursor inside a horizontal ruler to click and then drag a horizontal guide.

Guides are non-printable, so use them as needed to create order within your documents. Guides work well for developing templates that you can later share with colleagues or clients to help speed along the design process while ensuring accuracy. You can use the *Ctrl/Command + ;* keyboard shortcut to quickly view (and hide) your guides:

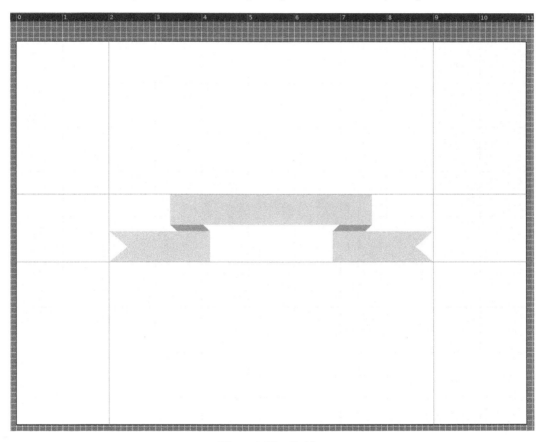

Figure 1.17 – Guides

As mentioned earlier, the guides can be created by pulling them from the rulers. After they have been created, they can be moved by dragging them, or even deleted (just drag them back into the ruler). Go to the menu bar and choose **View** > **Guides** > **Lock Guides** to keep from accidentally moving guides around or protect them as a template before sharing with others. To make your guides more visible, open your system preferences (remember that this was discussed earlier in this chapter) and change the color of your guides to something that offers a greater contrast to your work:

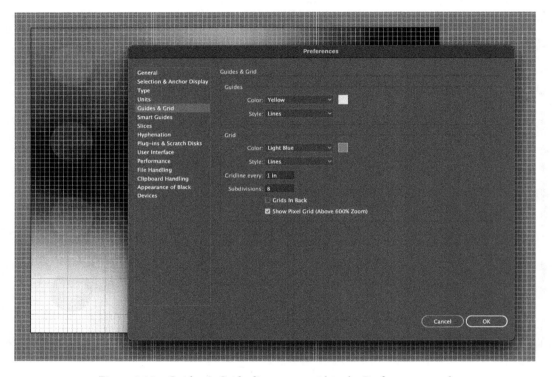

Figure 1.18 – Guides & Grid adjustments within the Preferences panel

In *Figure 1.18*, you can see that the work is predominately made up of cool colors, so having guides with a color with a warm hue helps to separate it from the artwork and offers greater visibility.

Consistency will be the key to mastering your designs, and using these layout aides should allow you to check for accuracy and deliver work that you are confident in its exactness.

As much as we all try to stay on top of things when it comes to the technology we use, it is inevitable that upgrades and improvements (even setbacks, at times) will continually require us to learn more. Even if you are comfortable with Adobe Illustrator and the topics I have addressed so far, you may find this book helpful in discovering some of the newest tricks this tool can do.

What's new in Adobe Illustrator 26.3.1?

Adobe is continually improving its software and releasing updates, so the following information relates to the May 2022 release of Illustrator on the desktop (version 26.0). We will be going over many of these items later in the book, but for now, I'll just offer a quick highlight of what each is and what it can help you do. In addition, you may want to select the **What's new** button, which is located at the bottom left of the **Home** screen, for the very latest additions.

Adobe Sensei

With Adobe's ongoing quest for improvement, they continually advance their software to the cutting edge of technology. Nothing can better illustrate that than their latest efforts in **artificial intelligence (AI)**. **Adobe Sensei** uses a neural network to make informed decisions. Adobe Sensei will help you create and deliver content with less effort or fewer steps. This is an ongoing process of software engineering, and updates are continually bringing us more opportunities for success. To view their latest offerings, be sure to visit `https://www.adobe.com/sensei.html`.

Illustrator on desktop

The following list of features are all examples of what Sensei adds to Adobe Illustrator:

- **Puppet warp**: Creates a warp mesh over your art, which can then have adjustment pins added for controlling the warping effect.

- **Global editing**: Similar items can now be intuitively selected and batch edited.

- **Curvature tool**: This uses past usage and experience to make an educated guess about the angle and distance of the curves created by the tool.

- **Freeform gradients**: This is an advancement from the mesh gradients that you may be used to. I think you will find this much easier and more logical.

- **Content-Aware Crop**: Illustrator can make crop suggestions based on the image and your intended output.

- **Recolor Artwork**: The Recolor Artwork panel has just gone through a major update and now allows for more color variations. It is now more compact and scaled down to just show the more popular tasks, but still offers an **Advanced Options** button to get to the panel as it used to be.

Aside from Sensei, Adobe has also introduced several additional features to Illustrator in this latest update:

- **Collaborative editing**: Edit with a friend or colleague by utilizing Adobe Cloud sharing.

- **Patterns in a click**: **Repeat** is Illustrator's new way to produce complex patterns quickly. Under the **Repeat** command, you can now develop patterns using either **Radial**, **Grid**, or **Mirror** methods.

- **Enhanced Glyph Snapping**: Aligning your graphic elements to your text in several ways.

- **Cloud documents**: Adobe has made major strides in developing a true cloud-based option for file storage and integration. Although they have used the Creative Cloud moniker for many years, it is only recently allowing full-scale storage, offline editing, and collaboration.

In addition to all these latest features that have been brought to the desktop version of Illustrator, Adobe has also been working hard to improve Illustrator on the iPad.

Illustrator on the iPad

Adobe has already had success with adapting several of its desktop applications to mobile devices, and now it is Illustrator's turn. At this time, they have developed the Illustrator iPad version, but have not developed a version for other platforms. If history repeats itself, they may develop something for additional platforms later.

Here are some new features for Illustrator on the iPad:

- **Work offline**: This is welcome news to anyone that travels and does not always have the good fortune to be connected to the internet.

- **Advanced export controls**: More options for file types and control for each have been added. You can export your iPad illustration as a JPEG, PNG, SVG, PDF, or PSD. Depending on which file format you choose, several options will be made available to you, such as color model, quality, and the resolution intended for the document. In addition, you will now be able to export the entire document, an artboard or range of artboards, or a selection.

- **Enhanced Blob Brush**: This tool is now working much like it would in the desktop version.

- **Transparency grid viewer**: Bringing over this display option from the desktop version, you can now set the view of your artboard to transparent instead of the default white.

- **Support for template format**: Adobe has taken a major step toward making the app a professional-level tool by adding support for template format. This allows the use of prefabricated templates to accelerate your workflow.

> **Important Note**
>
> Adobe Illustrator on the iPad is getting updates at a feverish pace, but this book will cover the options that are available at the time of publishing.
>
> At the time of this writing, Adobe has updated Illustrator to include native support for the new Apple Silicon M1 chip. This is a perfect example of how the improvements come at a faster and less regulated pace than in the past. Since changing from a single purchase (with upgrades) model to the subscription model, we have been given access to updates as soon as they have been beta-tested and deemed ready for consumers.

Summary

After having completed this chapter, you should now have the base knowledge needed to begin delving deeper into the options, enhancements, and recent additions to Adobe Illustrator so that you can truly master the software and have greater success in bidding it to do what you wish. This level of control falls into five key categories; they are workflow, customization, control, efficiency, and organization. Under the category of workflow, you should now understand how to find an appropriate workspace for your needs and save customized workspaces for more specific situations. Under the categories of customization and control, you should be able to find the Properties panel to customize Illustrator and the Control panel to select attributes and options for a selection. Under the category of efficiency, you should be able to accelerate your work with the use of panels and shortcuts for often-used tools and tasks. And finally, under the category of organization, you should know how you could stay more organized by using rulers, a grid, and/or guides.

In the next chapter, we will be preparing Illustrator for the instruction, activities, and challenges throughout the rest of the book. After prepping the software and discussing some suggested methods of digital asset management, you should feel comfortable and confident that you can move on and grow in both your knowledge and skill.

2
Prepping for Illustrator

Although it isn't overly complex to get started in Illustrator, there are several things you can do to improve your results. To begin a file, it's as easy as just hitting the **Create new** button on the left panel of the software's home screen. Once you do that, the **New Document** screen pops up with several additional choices for you as you prepare your document.

In this chapter, we are going to discuss developing a plan, consideration for output media, and utilizing a logical **digital asset management** (**DAM**) plan for keeping track of all your work and keeping your computer running fast.

To accomplish this, the chapter will be divided into the following main topics:

- Developing a plan
- Utilizing resources
- Media categories
- Artboards and Bleed
- RAM, scratch disks, and GPU performance
- DAM plan for organization and computer maintenance

Technical requirements

To complete this chapter, you will need the following:

- A sketchbook and drawing materials of your choice
- Adobe Illustrator 2022 (version 26.0 or above)
- High-quality internet access may be required for some situations

Developing a plan

A crucial part of beginning any task is having a plan in mind. When struggling to create in digital media, you should consider stepping aside from technology and developing the concept first, whether that be in the form of sketches or research (or a combination of both).

Sketchbooks are an ideal tool for starting your creative workflow. They allow you to fluidly conceptualize your design and make a multitude of variations with little time lost. They also allow you to quickly get your ideas down on paper and then you can later transfer them into Illustrator for further editing.

Illustrator allows you to bring in any raster-based image you want to work with. After importing the image into the desired layer, you can change it into a template layer, and it will automatically lock and dim the layer. In the **Layers** custom settings, you can dim the image to whatever works best for your specific situation. Note the change in the layer icon for the bottom layer that has been changed to a **Template** layer using the **Layers** panel's contextual menu:

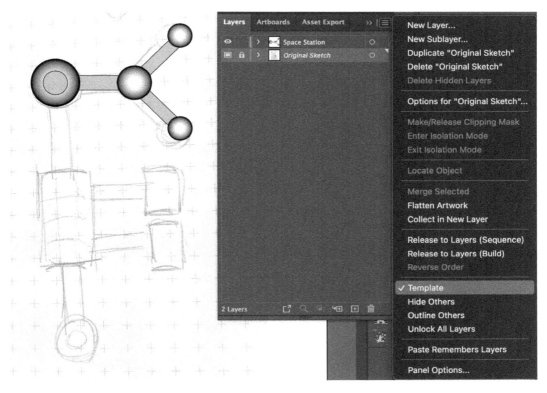

Figure 2.1 – Creating a template layer from a sketch

After changing the layer to **Template**, you can now further adjust it in the following ways:

- Under the **Layers** panel's contextual menu, you can choose **Options for "name of layer"** to find several options for layers, including **Dim Images to** for any **Template** layer. This allows you to soften the contrast of the image, giving it the appearance of tracing paper, but it does not change its opacity. Since the **Template** layer remains 100% opaque, you will need it to be the bottom layer.

- Turn on **Grid** using either the **Properties** panel or *Ctrl/Command + '* (you may choose to uncheck **Grids in Back** in the **Preferences** panel to allow it to be on top of all layers).

- Turn on **Rulers** using either the **Properties** panel or *Ctrl/Command + R*.

- Turn on **Guides** with either the **Properties** panel or *Ctrl/Command + ;* (semicolon).

- Turn off the **Template** view with *Shift + Ctrl/Command + W*.

It may feel a bit like tracing paper, but remember you are actually creating your vector work on top of the **Template** layer.

The key point I want you to take away from this is that great ideas come from *you* and not the technology. Sketching allows you to process a concept (and its iterations) much faster than going directly to Illustrator. It often allows a more creative idea to come out because you aren't fixated on how to do something in the software, but instead, are focused on what you would like the final output to look like. Of course, the more you look at good designs, the more you will possess good ideas for what you want your work to look like.

Utilizing resources

Another way to feel more prepared to use Illustrator and continually keep your work at its peak level is to keep an eye out for techniques and trends of other creatives. There are a lot of resources out there to choose from that allow you to be inspired. You can still be creative and original while using these resources to learn from other artists.

The following is a short list of a few I would highly recommend:

- Adobe's portfolio site, Behance (`https://www.behance.net`):

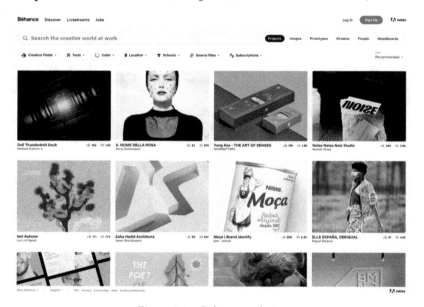

Figure 2.2 – Behance website

- Another excellent portfolio site, Dribbble (`https://dribbble.com`):

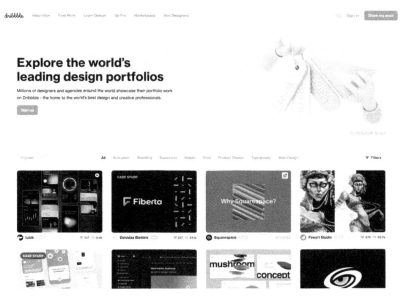

Figure 2.3 – Dribbble website

- An essential book for logo designers, *Logo Modernism*, by Jens Muller:

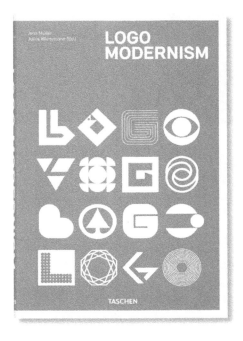

Figure 2.4 – Logo Modernism, by Jens Muller

- A book chock-full of design inspiration, *Pretty Much Everything*, by Aaron Draplin:

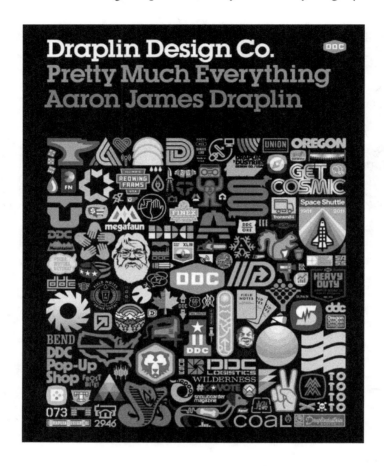

Figure 2.5 – Pretty Much Everything, by Aaron Draplin

- A subscription site for iconography, The Noun Project
  (`https://thenounproject.com`):

Figure 2.6 – The Noun Project website

- An excellent set of plugins, Astute Graphics (`https://astutegraphics.com`):

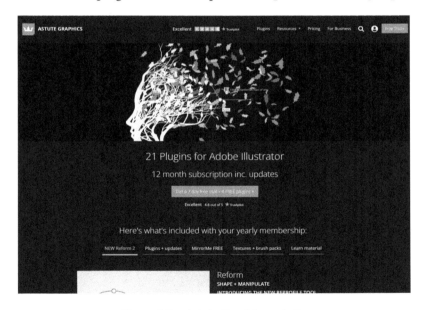

Figure 2.7 – Astute Graphics website

Of course, there are countless additional sources for inspiration and training: YouTube, Skillshare, workshops, podcasts, books, websites, and more! Just remember to enjoy the preparation as much as the creation and see how much your work improves. Once you feel prepared, it's time to choose which type of media you'll need for your file.

Media types

Once you have initiated Illustrator, you will be greeted with a welcome screen, which includes the following categories at the top of the **New Document** screen:

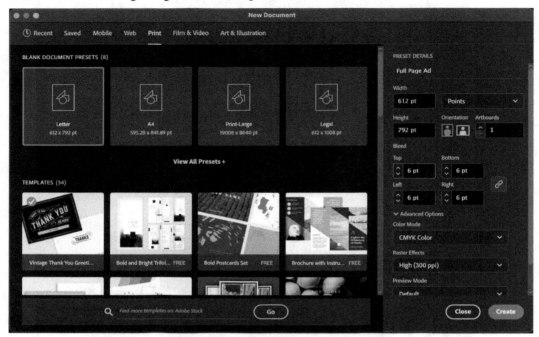

Figure 2.8 – The New Document screen

The **New Document** screen keeps track of any recent files you have been working on as well as any templates you have downloaded from Adobe Stock. Aside from the **Recent** and **Saved** categories, there are five additional media categories from which you can choose: **Mobile**, **Web**, **Print**, **Film & Video**, and **Art & Illustration**. Each of these media categories offers you several preformatted templates, each of which has their own specific purpose. Generally, these fall into two camps: items for print, and items for screen output. There is an exception though. In both the **Print** and the **Art & Illustration** categories, you will be offered several templates that offer common print sizes, but the **Art & Illustration** category is often set up for better screen viewing. Of course, you can adjust any of the document settings after they have been created if your or your client's intent for it has changed.

Although you will still be able to make further adjustments within the software, it is good practice to consider the output of your file as you are creating it. It develops good work habits where less needs to be decided when preparing to distribute the file for its ultimate purpose. If the intent of the file hasn't changed (that is, it is still intended for print), then simply saving the file with a keyboard shortcut of *Shift + Ctrl/Command + S* is all you need. This will bring up the **Save As** dialog screen, where you can choose the location you would like the file saved to.

Depending on the intent and/or media type, the next two options – **Artboards**, then **Bleed** – may prove to be very helpful tools available to you in your document.

Artboards

Artboards retain your art to the chosen media's resolution or proportion. With the ability to have multiple artboards, we have additional leverages. **Artboards** can be used in a multitude of ways, but let's go over a few popular methods for their use:

- View options
- Multiple artboards within one file
- Combination of different-sized artboards in a single file
- Combination of **Landscape** and **Portrait** mode artboards in a single file

Let's take a look at each method.

View options

Let's go over the view options first. After selecting **Artboard Tool** (*Shift + O*) from your toolbar, you will find that the **Properties** panel now presents an **Artboard Options** button to you under **Quick Actions**.

This will allow you to add a couple of valuable guides to your artboard view. Under the **Display** section of the panel, you can add a check mark to **Show Center Mark, Show Cross Hairs**, and **Show Video Safe Areas** (if you are working on video content). The Center Mark and Cross Hairs options allow you to better align content, as you will now see these elements continually on your artboard as a faint green guide, as shown here:

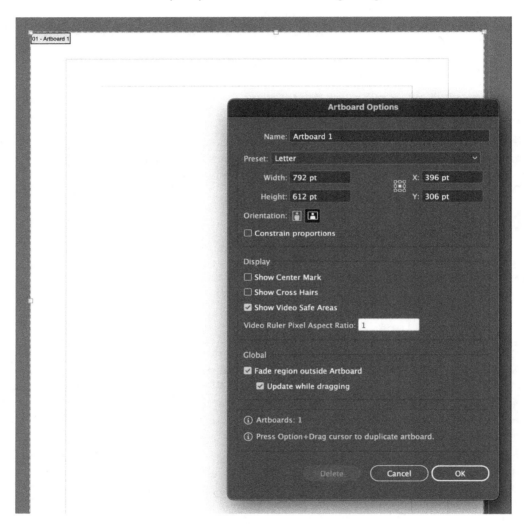

Figure 2.9 – View options for artboards

Although the Video Safe Areas were developed for TV broadcasts that would be presented on tube televisions, those devices are now few and far between. However, it is still recommended for video content creation, as this content is often adjusted and stretched from 4:3 to varied HD resolutions, and some content may be affected negatively.

If you are unfamiliar with what these guides represent, they represent three areas and their intended content:

- The innermost box is the title-safe area.
- The next box is the action-safe area.
- The area between the border of the action-safe area and the border of the artboard is the overscan (invisible) area.

These guides are represented as green lines in the following figure:

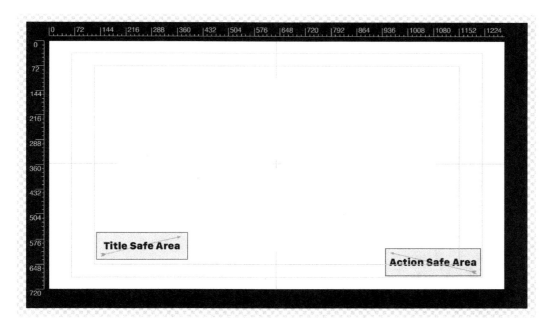

Figure 2.10 – Video Safe Area view option for artboards

The overscan area was given the nickname "invisible area" because, although it would indeed be part of the broadcast transmission, tube TVs would not be able to show that part of the broadcast. The action-safe area was thought to be the part of the video that most screens would be able to present to their viewers, but the title-safe area was used so that all TVs would be able to present text content without any clipping in this area. This was because the different manufacturers did not create their tube televisions to the same standard. Today's flat-screen televisions can present near edge-to-edge video from the television transmission, but it is still recommended that you consider the safe areas. This is even more important if you are making the content for streaming, as the device and/or end user may adjust the video player from its original aspect ratio and scale.

Multiple artboards within one file

You can think of this option as having multiple documents inside one file. To create additional artboards, you can use the **Properties** panel (be sure that no object(s) are selected). You should see an **Edit Artboards** button that will then adjust the **Properties** panel and allow for more artboard customization.

Here, you should see a group of four icons: the **New Artboard** icon, the **Delete Artboard** icon, and the **Portrait** and **Landscape** icons. Directly to their left, you will find that you can also choose a preset and name for the currently selected artboard.

Selecting the **New Artboard** button is one of the quickest ways to add an additional blank artboard to your current file. You can also tell Illustrator to create multiple boards when you are first setting up your document, under the **Create New** prompt.

You can view the artboards in the **Artboards** panel; it is often located beside the **Layers** panel, but if it isn't available, you can always find it (or any panel) alphabetically under the **Window** menu at the top of the interface. One benefit of having multiple artboards is that it allows you to host variations of a design within one file. The following screenshot shows the panel's location:

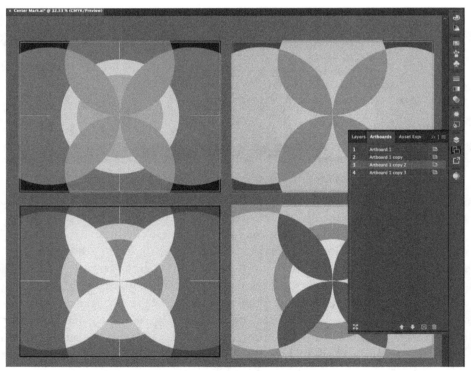

Figure 2.11 – Multiple artboards

You can navigate between artboards by selecting them in the **Artboards** panel and double-clicking on any area except for the name (as Illustrator will think you are wanting to rename it). It will instantly zoom in to fit the chosen artboard to the window, as shown in the next screenshot:

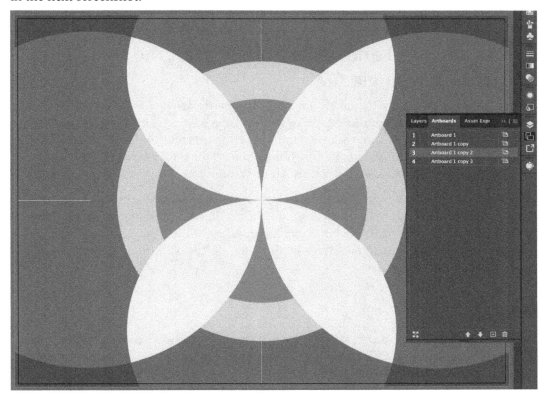

Figure 2.12 – Selecting a single artboard view

In the example illustrated in *Figure 2.11* and *Figure 2.12*, you would be able to jump from one artboard to the other by double-clicking on the one you would like to target, making adjustments (such as color), and then double-clicking on another one to edit it next.

Combination of different-sized artboards in a single file

Another great benefit to artboards is the ability to allow them to present different media. After selecting one of your artboards, you can change its purpose by choosing a new preset from the list. There are presets for print, web, video, and devices, with several popular sizes for watches, phones, and tablets available. In addition, you can create a custom size if none of the presets work for your intent.

Illustrator allows you to host not only multiple artboards but multiple *kinds* of artboards within one file. An excellent example of the use of this technique is in creating an entire campaign for a brand design in one file. Designers can create one artboard for the letterhead as well as one for a business card and another for an envelope. Using this approach allows all the designs to be kept together within one file.

Combination of Landscape and Portrait mode artboards in a single file

Another benefit is the ability to quickly adjust the artboard's orientation to either Portrait or Landscape view. One application it works well for is presenting orientation on devices if you are a UI developer. If you work with multiple artboards often, you are going to love the following two keyboard shortcuts: *Ctrl/Command + 0* for **Fit Artboard in Window,** and *Ctrl/Command + Alt/Option + 0* for **Fit All in Window.**

The ability to have your files hold multiple artboards that can allow you to stay more organized and efficient has been a welcome addition to Illustrator for many users. I think you will find many ways you like using it too. It may take a while to get used to all the benefits of artboards, but after you have used them for a while, you will never want to be without them.

Bleeds

Bleeds are necessary (and usually required) when sending your work to a printing service and you want to be sure that all artwork will always make it to the edge of the document. This small area of Bleed is where the printer can account for small adjustments to the cutting and still be sure your work doesn't end up with any unattractive white edges. The Bleed space can be created either when beginning the document or can be adjusted from within the **Properties** panel. When nothing is selected, you will see the **Document Setup** button under the **Quick Actions** area of the panel. Select this, and you will then be able to add the desired dimensions for the Bleed:

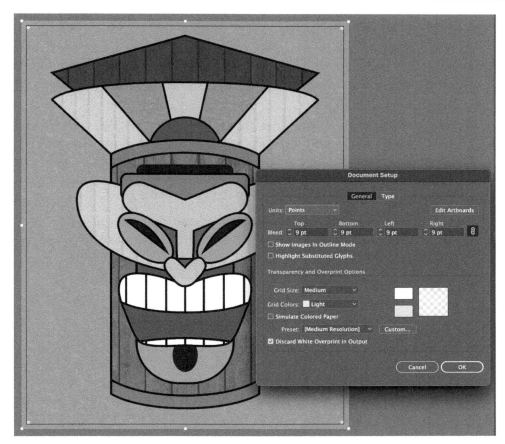

Figure 2.13 – Adjusting Bleed inside the Document Setup panel

After entering the **Bleed** amount, be sure to extend your artwork to the new **Bleed** guidelines, as in *Figure 2.13*.

When you are preparing to deliver your file to the printer, you want to save it as a PDF and check **Use Document Bleed Settings** in the **Marks and Bleeds** options panel. If you did not yet set them up in the document, then you can add them on this same panel, but be sure you have extended your artwork to this Bleed distance. Otherwise, you will just be creating a white border around your work. It is recommended that you talk to your printing provider to determine the Bleed size they would like.

Knowing how to set up Bleeds will assure you of a quality print within the tolerances given to you by your printing provider.

Throughout this chapter, the theme has been saving time by planning for what might be ahead. An important aspect of time management is having hardware that is sufficient for the job and has been well maintained for dependability over time.

RAM, scratch disks, and GPU performance

Having adequate **random-access memory (RAM)** is crucial to keeping Adobe Illustrator from processing slowly or even hanging during a complex command. Adobe states that the minimum system requirement is 8 GB but recommends 16 GB or more.

In addition, you should try to keep plenty of hard drive space available on your **scratch disk**. The scratch disk needs to be a local drive. This is usually your startup drive because it should be your fastest drive. A **solid-state drive (SSD)** makes for a great scratch drive since it is so fast. Before beginning, just direct Illustrator to your scratch disk using the **Preferences** panel. If you have more than one hard drive installed, you can also direct Illustrator to a secondary scratch disk from within the **Preferences** panel.

The third item that can really accelerate your workflow is the computer's **graphics processing unit (GPU)**. This is often known as your **graphics card**, and Adobe lists several that are powerful enough to support the GPU performance features that are offered in Illustrator. To see a current list of these devices, go to the following link: `https://helpx.adobe.com/illustrator/system-requirements.html`. Generally, we need something that has more than 2 GB of **video RAM (VRAM)**, but again, it seems that the more the better.

It is very important to check your hardware to make sure that it will allow Illustrator to work for you as intended. From there, you can focus your energy on planning, creating, and organizing your work. Staying organized is difficult, but it offers benefits to you and your hardware over time that you shouldn't ignore. Having an organized system improves your odds of finding what you're looking for, as well as keeping your hardware running Illustrator at peak performance level.

Digital asset management plan for organization and computer maintenance

DAM is the process of utilizing a logical, organized, and repeatable method of file control that allows the user to be more efficient. It should also allow you to have a much greater likelihood of finding what you are going to be looking for in the future. There are many ways to do this, and it is often not achieved without some trial and error.

The method I have adopted and recommend to others is the "time capsule" method. I start by building folders for a current year and then inside, place a folder for each month. Some people like to stay even more organized by creating folders for weeks or even each day, but I find that to be too specific. I have less to remember if I just bring it down to the month and year that the file was made. From inside the month folder, I usually build client or project folders (a client folder often gets project folders within it). If the date is not your most important factor when searching for files, you may have your client folders at the top level and then include the date, and then project folders inside. If you build a pattern that you continue to repeat, you should find it easier to manage files over time. Of course, you can allow technology to do search queries, but those do not always present the results we hope for.

Where you store all your files is also an important consideration. I highly recommend you store the bulk of your work on an attached drive to reduce the space used on your computer's main hard drive.

I find that developing your DAM plan is a continually evolving process and may never be perfect. Over time, though, I'm sure you will find the process easier and well worth the effort.

Summary

After having completed this chapter, you should now feel more confident in your preparation to begin creation. You have learned that there are some distinct benefits to planning and preparing your vision before beginning in Illustrator, as well as how to prepare the document once you do get started. I have aimed to emphasize why this is important to the creative process. Sketches and research should add quality and richness to your concepts before even entering the digital tool we know as Illustrator. It's important to remember that Illustrator is just that, a tool. Like any other tool, it takes training and practice to master.

I have also given you a bit of advice about organization in this chapter, whether it is organization within a file using templates and artboards, or the bigger picture of all your files being organized into a DAM workflow. The time it takes initially should help you avoid far more wasted time and stress later.

In the next chapter, we will be moving into the first steps of how we can create and manage objects with Illustrator. We will review how multiple objects can be collected into a group and utilized to create a compound path or a clipping mask. We will also be discussing the use of layers and migrating objects from one layer to another.

Part 2 – Advanced Illustrator Methods

This is the main part of the book and offers strategies for progressing the basic skills you have already acquired from prior experience.

This part comprises the following chapters:

- *Chapter 3, Developing and Organizing Objects*
- *Chapter 4, Drawing with the Pencil, Paintbrush, Pen, and Shape Tools*
- *Chapter 5, Editing and Transforming Objects*
- *Chapter 6, Advanced Attribute Design*
- *Chapter 7, Powerful Typography Options in Adobe Illustrator*

3
Developing and Organizing Objects

Now that we have some of the basics out of the way, we will start discussing the process of creating files within Illustrator by addressing basic organization with object and layer management techniques. This may seem a bit out of order, as we won't discuss creating objects in detail until the next chapter, but knowing a variety of organizational techniques will help you maintain a good level of control once you start creating. After completing this chapter, you will know how to select and combine objects, create layers, and migrate items from one layer to another.

Whether you are new to Illustrator or feel fairly accomplished in it, this chapter aims to encourage you to learn about and/or revisit methods of file management that will help you be more efficient with your time.

To accomplish this, the chapter will be divided into the following main topics:

- Groups and compound paths
- Layers and selecting
- Migrating objects into other layers or sublayers

Technical requirements

To complete this chapter, you will need the following:

- Adobe Illustrator 2022 (version 26.0 or above).
- High-quality internet access may be required for some situations.

Groups and compound paths

Illustrator allows you to hold together a collection of objects in a **group**. Groups allow you to gain a lot of control over your work by reducing what is, at times, a seemingly endless list of vector objects in a layer.

Groups work well for most situations, as they allow for much greater organization. Occasionally though, you will have a situation that calls for a special relationship between two or more objects. That's when the next method can be utilized.

Compound paths feel like groups because they take two or more objects into consideration, as they use the top-level objects to cut holes in the one below. The technique is limited in its uses, but some examples where it would work well include glasses, windows, and rings.

We will examine both of these techniques in greater detail and discuss how you employ them to take greater control of your work.

Groups

I consider the use of groups to be a necessity to avoid chaos. By design, Illustrator will continue to add objects into the current layer indefinitely. If you are coming from having more experience in Adobe Photoshop, then this will be something to get used to. Many people who are new to Illustrator find this quite aggravating, as Photoshop seems to do a better job of assisting you with layer creation.

Each time you create a vector object in Photoshop, a new layer is created. In Illustrator, the responsibility is yours to develop a method of organization or not. If you choose to do nothing, Illustrator will happily continue to drop every solitary object into the active layer. From my experience, you will likely regret it if you choose to let this continue for very long.

The following are steps to make groups (and later, **layers**) work for you:

1. Continue drawing several shape objects and apply the desired attributes (**Fill** and **Stroke**, for example).

2. Any time you have developed enough to form a recognizable group that may later need to be collectively changed (scaled or moved, for example), then select all the individual objects that will make up the group and then choose **Object** > **Group** from the **Menu Bar**, or you may use the keyboard shortcut of *Ctrl/Command + G*. Another method is to right-click and then choose **Group** from the contextual menu.

3. Continue this process for additional recognizable items. For example, imagine that you are designing a face; you might create one group for all the objects that make up the right eye and one group for all those that make up the left eye of that same face. In addition, you could then add those two groups into another group, such as an *eye* group, if you would like to edit them together.

4. As you continue, you can go on to collect groups to put into an even broader grouping, such as *face*.

> **Important Note**
> Selecting an object is as simple as clicking on the object's vector path. To select multiple objects, just hold the *Shift* key while clicking on multiple objects' vector paths. In addition, you may find it easier to *Shift*-click on the small spaces to the far right on the **Layers** panel.

In *Figure 3.1*, you can see that, without some method of organization, the **Layers** panel will quickly become a mess of unlabeled objects. To make things even more difficult, they are not even to scale since it shows only the actual object in the thumbnail view:

Figure 3.1 – Layer with ungrouped objects

The small blue boxes to the right of each object in the **Layers** panel indicate that they are all selected. Select all the objects you intend to add to a group.

After selecting all intended objects and then using the shortcut of *Ctrl/Command + G*, the layer should appear, as in *Figure 3.2*:

Figure 3.2 – Layer with grouped objects

Utilizing groups will allow you to stay more organized and have greater control of your work. Another benefit is the acquisition of additional space within your **Layers** panel. Unless you're a big fan of scrolling, groups are sure to bring you some added sanity in your workflow.

Another method of collecting objects that I use quite often is to lock those that you do not intend to add to the grouping. The small box to the right of the **Visibility** icon (the eye) is the **Lock** icon (the padlock), as shown in the following figure:

Figure 3.3 – Enlarged view of layer locks

By selecting this on every object or layer that you do not intend to select, it will no longer allow them to get selected. Then, you can choose **Select** > **All** from **Menu Bar**, or you may use the keyboard shortcut of *Ctrl/Command + A*. This allows all available objects to be selected at once. It's a real time saver compared to *Shift*-clicking all of them.

You can lock groups within a layer, or an entire layer can be locked. To lock certain objects or groups within a layer, you must first toggle open the layer they reside in.
You can do that by clicking on the right-facing angle bracket located to the left of the layer's name.

Here are a few additional points for managing your objects, groups, and layers:

- Rename them by double-clicking on their current name and then customizing it with something that will help you recognize them as you continue working with the file. If you click in the empty area next to their name, a **Layer Options** panel will open, and you can also enter the new name at the top of this panel.

- Recognize that a layer may appear to be unlocked (no lock icon is visible), yet one or more objects or groups will not allow you to select it on the artboard. Toggle open the layer to see whether anything contained in the layer has the lock icon visible.

- Use *Shift + Ctrl/Command + G* to ungroup any current grouping, and then select the part that you would like to edit.

Grouping, ungrouping, and then grouping again will become a common habit for you, if it isn't already part of your design process. That technique, along with renaming objects, groups, and layers, will keep your file from quickly getting out of your control, and finding what you're looking for will be much simpler.

Another method of combining two or more objects is the compound path method, although it differs from a group in one distinct way, which we will review next.

Compound paths

A **compound path**, like a group, consists of two or more objects, but the difference is that these objects interact with each other when put into a compound path. They are used when you would like to show an underlying object through a hole in another object. To illustrate this, I have added some circles to the Tiki file you have seen throughout this chapter. You can see in the following screenshot that I have given them all a light blue fill:

Figure 3.4 – Multiple objects placed on an underlying object

To use those four light blue circles in a compound path, they must all be selected as well as the objects directly below them. This needs to be done in two parts.

First, the triangular object and the three circles above it must be selected, and then the semi-circular object and the circle above it can be processed. To create the compound path, make sure all the associated objects are selected, then right-click on them to select **Make Compound Path** from the contextual menu. The result will be an object with one or more holes in it, as in the following screenshot:

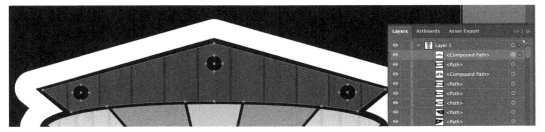

Figure 3.5 – Completed compound path

If you have used Illustrator for a while, you may be thinking that this looks like other techniques you might use. This is true, but a benefit to this approach is that you have not deleted the areas that are now represented as holes. If you right-click on the current compound path, you can select **Release Compound Path** from the contextual menu, and the objects that have been creating holes will once again be objects with their own properties. They will, though, default to the attribute of the underlying layer. This is illustrated in the following screenshot:

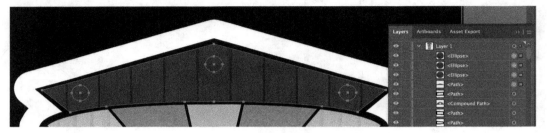

Figure 3.6 – Released compound path

Of course, these newly released objects can be reselected and adjusted as needed. Choosing **Select** > **Deselect** from the top menu or using the keyboard shortcut of *Shift + Ctrl/Command + A* will allow you to deselect the current selections and then *Shift*-click the desired objects. Now, you can either choose their visual attributes or delete them entirely. Just remember that if they are deleted now, there is no longer a hole in the underlying layer so they will no longer have any visual presence.

Compound paths don't have as many uses as a group would, but they work well for building things that may need a hole, while still having the ability to revert to an object that will have a filled opacity. Things this could work well for include glasses, windows, and donuts.

Similar to the compound path, there is an option to create a **compound shape**. The compound shape, like a compound path, allows you to connect two or more items but allow you to make adjustments to each. We will review methods of creating compound shapes in *Chapter 4, Drawing with the Pencil, Paintbrush, Pen, and Shape Tools*, as well as the method for easy editing within them, which is known as **Isolation Mode**.

Now that we have discussed groups and compound paths, you can create very advanced and complex illustrations, while utilizing a tried-and-true method of simplifying them down to an easy-to-handle grouping. Next, we will explore the benefit of using layers to gain even greater control of your creations.

Layers and selecting

An important lesson that all beginner Illustrator users learn is the need to keep the **Layers** panel open. Objects that you see on your artboard rarely tell the entire story, and the **Layers** panel is your tool for looking into the construction of your illustration. It helps you see where each part resides and how it is being presented. Think of it as your blueprint for the design, and you will begin to develop your work with the understanding that it is a construction that you are engineering.

Taking this analogy even further, a well-built design will hold up well, while a poorly constructed or organized design may be difficult to maintain and even have the potential to crumble as it gets developed. This is even more true if you are collaborating with others, as all the team members must be able to understand the structure being used.

As we discuss the use of layers and techniques for selecting, you may want to consider them to be like folders in a file cabinet. They can be added or removed at any time and can each hold a collection of items. As we take a closer look, remember that there isn't just one correct way to use layers; it ultimately comes down to the needs of the user.

Layers

A further step in getting organized within Illustrator is in the utilization of **layers**. As previously pointed out, without your intervention, Illustrator is happy to continue loading up a single layer in perpetuity.

To create a layer, you must have the **Layers** panel visible. Choose **Window** > **Layers** from the top menu if you do not currently have it open. Note that all visible items in Illustrator are part of a stack; as items are created, they will be added on top of previous items.

In the sample illustration shown in *Figure 3.7*, you can see the designs for a badge – **Layer 1** shows a car, while **Layer 2** shows an additional car with a variation to the design. Since both layers are currently visible, you would only see the car in **Layer 2**, as it completely overlaps the car in **Layer 1**.

To create another layer above both of these layers, you can either select **New Layer** from the upper-right contextual menu or select the + sign just to the left of the lower-right trash can icon, as shown here:

Figure 3.7 – Creating layers

Once you have added a new layer, this new layer will allow you to create more content inside it or migrate existing content to it.

This is also a great time to consider renaming the layer. Renaming the layer is not necessary for the function of the file, but I would highly recommend you consider it for your own sanity. It makes it far easier to find what you are looking for. To do this, you can just double-click on the current layer name, type in the desired name, and hit the *Enter/Return* key. You can also select the layer and then choose **Options for "name of layer"** inside the **Layers** panel's contextual menu. Yet another method for bringing up the same dialog box would be to double-click to the right of the current layer name.

Important Note

The **Options for "name of layer"** dialog box is also where you can adjust the layer indicator color. This is one more way that you can set up visual cues to assist you with making sense of more complex files.

Here are some techniques for using layers:

- Change the lock status of the layer by clicking the lock icon to lock or unlock the layer. This allows you to protect the entire content of each layer or sublayer.

- Change the visibility state of the layer between **Visible** and **Hidden** by clicking the eye icon to view or hide the layer.

- Change the visibility of the layer to **Outline** (a mode that shows only the vector path of the layer) by holding down the *Ctrl/Command* key while clicking the eye icon.

- To show/hide all other layers, hold down the *Alt/Option* key while clicking on the eye icon.

Other elements of the **Layers** panel also help you have greater control of your work, as shown here:

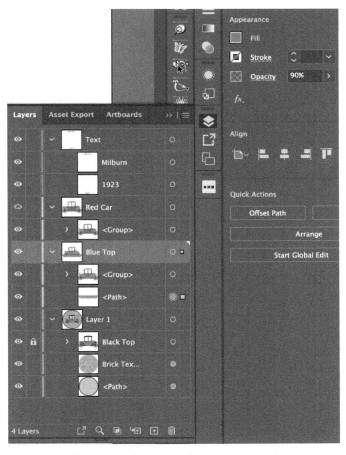

Figure 3.8 – Elements of the Layers panel

In the **Layers** panel, from left to right, *Figure 3.8* shows the Visibility column, the Edit column, thumbnail and layer name, the Target column, and the Selection column.

- In the Visibility column of *Figure 3.8*, you will see the eye icon, and you will notice that the **Red Car** layer is in **Outline mode**. This is symbolized by the outline of the eye icon.

- To the right of the Visibility column is the Edit column, where you can find the locks for each layer, and you will notice that the group that has the **Black Top** car illustration in **Layer 1** is locked. The entire layer is not locked, but rather just this specific sublayer.

- Next, you see the thumbnail of the contents of the layers and sublayers along with their name or type (layer, group, or object). The **Text**, **Red Car**, **Blue Top**, and **Layer 1** layers are also toggled open to show the sublayer contents. You can toggle a layer open or closed by clicking on the small arrow to the left of the thumbnail image.

- Next is the Target column and then finally, the Selection column. These two generally work together, but the one exception is that you can only target a layer in the **Layers** panel. The green box in the Selection column indicates that the path sublayer is selected. A selected element can then be engaged with in a multitude of ways. The double ring in the Target column indicates that the sublayer is also targeted for appearance attributes (note that in *Figure 3.8*, the targeted path shows in the **Properties** panel that it has a gray fill, no stroke, and is set to 90% opacity. The filled-in icon indicates that it already has appearance attributes beyond a single fill and stroke.

Let's take a closer look at how selecting works in the next section.

Selecting

There are two tools for selecting objects and groups: the **Selection** tool (*V*), which has the appearance of a black arrow, and the **Direct Selection** tool (*A*), which has the appearance of a white arrow. We will be discussing the **Selection** tool in this section, as it will allow you to select the entire object or group. The **Direct Selection** tool is for selecting and editing specific **Anchors** on your object's path. We will review methods for the use of the **Direct Selection** tool later in the book.

To select something in your illustration, click on it using the **Selection** tool. To select additional objects, hold *Shift* while clicking on other objects. If you are having trouble selecting multiple objects, I would recommend trying to click on the object's path, when possible. While selecting objects, you should look for the path's selection indicator color to understand whether it has been added to the selection. In addition, I would highly recommend keeping the **Layers** panel open to look at the Selection column for verification that an object is indeed selected. You can also add an object, group, or layer to a selection by clicking on it in the Target or Selection columns. Again, hold *Shift* if you are adding several objects to the selection. If a layer is in **Outline mode**, it will be ignored, and items in lower layers will instead be selected.

The paths for each item selected will be recolored to the layer's indicator color (or each anchor, when in **Outline mode**). This is very helpful in identifying what is and what is not selected, as shown here:

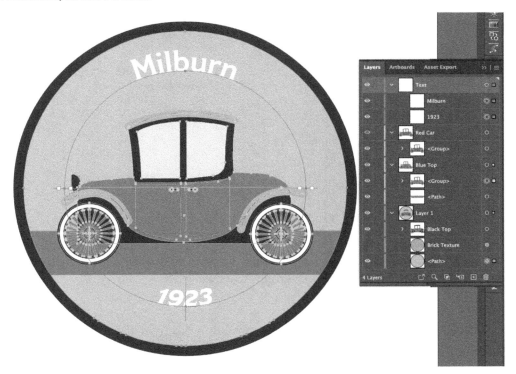

Figure 3.9 – Multiple items selected

As you can see in the previous screenshot, the entire **Text** layer is selected and is represented by the indicator color blue, the group inside the **Blue Top** layer is selected and is represented by the indicator color green, while the object of the badge shape in **Layer 1** is selected and is represented by the indicator color red.

There is a lot of flexibility when it comes to making your selections in Illustrator. After selecting any single object or a combination of objects, groups, and layers, you can make a multitude of changes, such as **Transform**, **Align**, and **Recolor Artwork**. These adjustments will be specifically discussed later in the book.

Next, we are going to discuss how you can adjust and organize the stacking order of your objects, groups, and layers. We will look at stacking these items within a layer, as well as migrating them from one layer to another.

Migrating objects into other layers or sublayers

Before discussing the process for migrating objects from one layer to another, we should discuss how you can go about arranging objects within a single layer. As I mentioned earlier in this chapter, Illustrator will continue to stack objects on top of one another, unless you intervene.

To make changes to the stacking order of objects, you can choose **Object** > **Arrange** from the top menu. From there, you will be presented with four choices:

- **Bring to Front**: *Shift + Ctrl/Command +]*
- **Bring Forward**: *Ctrl/Command +]*
- **Send Backward**: *Ctrl/Command + [*
- **Send to Back**: *Shift + Ctrl/Command + [*

You can use this method for anything within a layer, but it does not allow for migrating it from one layer to another.

Things work a little differently when moving something from one layer to another. Here are the steps for moving objects or groups to a new layer:

1. Make a selection.
2. Grab the selection icon from the Selection column (far right of the **Layers** panel).
3. Move it by dragging the object into the layer you want it to go to.

This will work for an object, a group, or an entire layer, but it will not work for multiple selections. If there are quite a few objects you would like to migrate to a new layer, I would recommend you first add them into a group and move them. Once in the new layer, the group can be ungrouped to allow you to continue editing the individual objects.

The skill of migrating is incredibly valuable, as you will inevitably design without layer development in mind. Having the power to organize and migrate your art to different layers allows you to focus on creativity first, then you can go back and create a more logical structure as the illustration grows.

As you get more familiar with your workflow in Adobe Illustrator, you will not only get more skilled with Illustrator's tools and effects but also in the best practices for organization. Through the methods of arranging objects, building groups and compound paths, creating and utilizing layers, and selecting and targeting objects, groups, and/or layers, you will start to tame your designs.

Summary

After having completed this chapter, you now know how to use groups and compound paths to compress complex illustrations into manageable components. You also have the knowledge to utilize and organize your Layers panel more successfully.

We have reviewed best practices for using the Layers panel and understanding all its columns. We also looked at techniques for moving items from one layer to another.

All the information within this chapter was intended to challenge you to work on the organization of your files and learn to reduce the long list of objects in your Layers panel.

In the next chapter, we will be discussing the variety of drawing and painting tools offered in Illustrator, as well as several techniques to improve your use of them.

4
Drawing with the Pencil, Paintbrush, Pen, and Shape Tools

Now, we are going to look at the variety of drawing and painting tools offered in Illustrator. There are a lot of options and approaches to each tool, so although this won't be a comprehensive list of everything you can do with each, I hope you find something helpful to add to your own bag of tricks after reading this chapter.

After completing this chapter, you should have expert knowledge of **paths**, **anchors**, **strokes**, and **fills**. You should also have a lot of additional knowledge on the use of **direction lines** and **direction points** for editing an object's path.

To accomplish this, the chapter will be divided into the following main topics:

- Making the **Pencil** tool work for you
- Loading and creating brush profiles
- Advanced training on the **Pen** tool and anchor point editing
- Exploring simple shape tools and the **Shaper** tool to develop more advanced objects
- The **Pathfinder** panel for heavy lifting

Technical requirements

To complete this chapter, you will need the following:

- Adobe Illustrator 2022 (version 26.0 or above).
- High-quality internet access may be required for some situations.
- Although not required, a quality graphics tablet or touch screen is recommended.

Making the Pencil tool work for you

To begin using the **Pencil** tool (*N*) (or any of the other tools we will be looking at in this chapter), we must first discuss what makes up all vector objects. All vector objects are made up of straight lines and/or curves. The curves are developed based on a mathematical formula. The benefit of this is that a vector path has smooth edges at any resolution, unlike a raster-based file, which is made up of a finite number of squares (also known as pixels) of color.

To that point, once you have selected the **Pencil** tool and its attributes, you will be able to draw freehand vector paths and closed vector shapes, as seen in the figure that follows:

Figure 4.1 – Artwork in Preview mode

In *Figure 4.1*, we see that the **Preview** mode allows you to see your creation with all its attributes applied to the vector paths. As we discussed in earlier chapters, attributes include options such as **Stroke**, **Fill**, and brush definitions. To see only the vector paths of the objects, choose the **Outline** mode – **View** > **Outline** or *Ctrl/Command* + *Y*.

The following figure shows how the art in *Figure 4.1* appears after completing the **Outline** mode command:

Figure 4.2 – Artwork in Outline mode

Viewing your artwork in Outline mode is an excellent way to focus on the quality of your shapes before enhancing them with attributes. The paths in *Figure 4.2* were created using the **Pencil** tool in combination with the **Smooth Tool**.

Using the **Pencil** may be a bit difficult at first, but the following suggestions will make it much easier to get this tool to do your bidding:

- Draw slowly and freely to create organic shapes. If you need a straight line, hold down the *Shift* key as you draw.

- As you get close to the start of your path, a small circle will appear next to the pencil icon to let you know that you are about to close the object. Just hover over this symbol and let go to close the shape.

- If any part of your path was not drawn to your liking, just redraw that section by beginning this path adjustment on the previous path, as well as ending it by connecting back to a previous part of the original path.

- If it makes a new path instead of correcting the previous one, just undo (*Ctrl/Command* + *Z*) and try again. Be careful to start and end on the original path.

hen you have a path you are generally happy with, use the **Smooth Tool** to smooth the path and reduce the anchor points along it. The **Smooth Tool** can be found in the same tool group as the **Pencil** tool or can be summoned by holding *Alt/Option* while using the **Pencil** tool. The **Pencil Tool Options** panel must have the **Option key toggles to Smooth Tool** option checked for this to work (see *Figure 4.3*).

Creating with the **Pencil** tool will take some time to get used to, but it is the most natural drawing method within Illustrator. If you find it difficult to make a long stroke using this tool, you can draw one section of the path at a time. When placing the tool close to the end of the current path, you should see a diagonal line beside the icon for the tool. This indicates the ability to connect to the previous path. Click and draw to connect and continue to add complexity to your path. The **Pencil Tool Options** panel must have **Edit selected paths** checked for this to work (see *Figure 4.3*).

You can also customize **Pencil Tool Options** by double-clicking on the tool's icon in the toolbar. You can also adjust the **Fidelity** slider to change the path from being more accurate to being smoother:

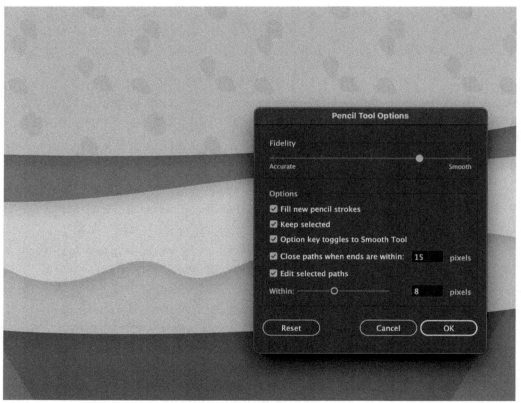

Figure 4.3 – Pencil Tool Options panel

The following figure highlights the **Pencil** tool's ability to create fluid, hand-drawn paths and shapes quickly:

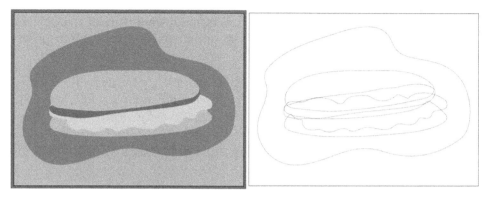

Figure 4.4 – Artwork created using the Pencil tool

The preceding artwork may initially have some odd forms, but these can easily be edited using the methods we have previously discussed in this section.

Drawing with the **Pencil** tool allows for a high level of flexibility in your work and can easily be edited using these redrawing and smoothing techniques. In *Figure 4.4*, you can view an early sketch of a sandwich in both **Preview** and **Outline** modes (use *Ctrl/ Command + Y* to switch between them), while in *Figure 4.5*, you can see that both the top and bottom buns have been redrawn by adjusting their active paths with the **Pencil** tool and **Smooth Tool**:

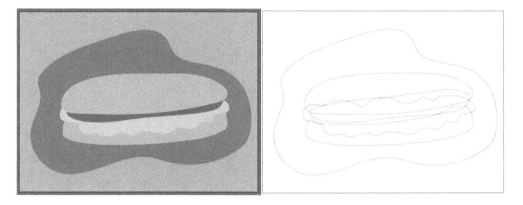

Figure 4.5 – Artwork edited using the Pencil tool redraw technique

After creating the path, just remember to be sure that it has an active path before trying to redraw over a portion of it with the **Pencil** tool again. If it is no longer active, you will need to use the **Selection** tool (*V*) to make it the active selection, and then choose the **Pencil** tool to start somewhere on the path as well as end the adjustment at a different area of the old path. This should allow you to try multiple times (if needed) to get the path you desire.

In addition, remember that you may also choose the **Smooth Tool** to allow Illustrator to reduce and adjust the anchors creating your object for smoother curve transitions. You can either find the tool by clicking and holding over the Pencil tool icon (you will find it directly under the **Pencil** tool icon) or you can just hold down the *Alt/Option* key while using the **Pencil** tool. This will temporarily change the **Pencil** tool to the **Smooth Tool** and then return to the Pencil once it is released.

Practicing these two techniques will allow you to adjust any vector path with ease. It will work with any previously drawn vector shape or path, no matter what tool you used to create it.

Next, we will be looking at customizing paths with a variety of **Paintbrush** tool (*B*) options.

Loading and creating brush profiles

The **Paintbrush** tool allows for a large variety of options. We will first look at how you can find those that are already installed in Illustrator, and then we will discuss how to create new ones.

Loading brush profiles

The default brushes can get you accustomed to using the tool and offer you several options, but eventually, you are going to want more choices for brush options. Knowing where you can find extra installed brushes is a good place to start when expanding your choices for customization:

1. Open the **Brushes** panel.
2. Access the **Brushes Library** menu in the upper right or lower left of this panel and choose a library.
3. In the new window, locate the desired brush profile.

In *Figure 4.6*, you can see what this should look like. When you locate the desired brush profile and click on it, it will be added to the document's current brushes. You should now see it as an option within the **Brushes** panel:

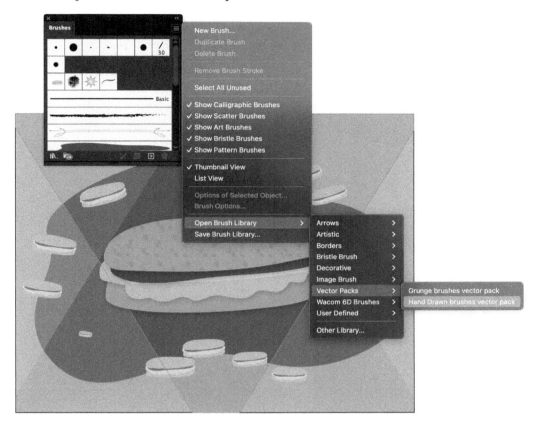

Figure 4.6 – Loading brush profiles from the Brushes panel

In addition to being able to load additional brushes, you can also adjust current brush profiles or build entirely new ones. Next, we will look at the options for brush profiles and how you can design your own custom profiles.

Creating brush profiles

There are five types of brush profiles in which you can create artistic strokes. These are **Calligraphic Brush**, **Scatter Brush**, **Art Brush**, **Bristle Brush**, and **Pattern Brush**.

The first of the five more traditional brush types we will look at is **Calligraphic Brush**. This is intended to create strokes that resemble marks made with the angled nib of a calligraphic pen. It uses a shape as a brush tip and routes it along a path, so a simple circle is used to create a non-calligraphic stroke, and changing the roundness and angle will result in a calligraphic stroke. In *Figure 4.7*, you can see the **Calligraphic Brush Options** panel. After opening the **Brushes** panel (**Window > Brushes**), double-clicking on a brush profile will bring up this options panel:

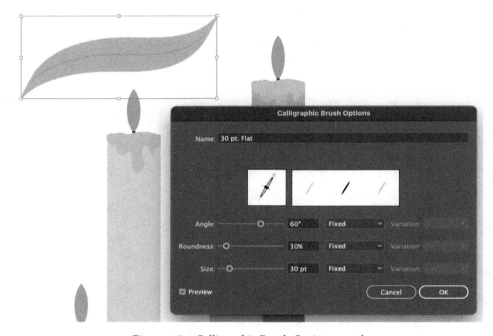

Figure 4.7 – Calligraphic Brush Options panel

The next brush to consider is **Scatter Brush**. It distributes copies of an image along a path. As you can see in *Figure 4.8*, you have several variable options with this style of brush. You can customize the brush's **Size**, **Spacing**, **Scatter**, and **Rotation** values. The seven options available for each of these variables are **Fixed**, **Random**, **Pressure**, **Stylus Wheel**, **Tilt**, **Bearing**, and **Rotation**:

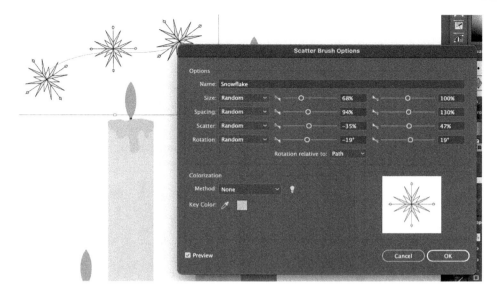

Figure 4.8 – Scatter Brush Options panel

One of the most often used variations of **Brush Tool** is **Art Brush**. This brush allows you to scale a single brush stroke shape onto a path. The options panel, as seen in *Figure 4.9*, allows you to choose between three scaling options. It may be scaled proportionately, stretched to fit the stroke length, or stretched between two predetermined guides that you adjust. The width adjustment can be **Fixed**, or adjusted at **Width Points**, **Pressure**, **Stylus Wheel**, **Tilt**, **Bearing**, or **Rotation**:

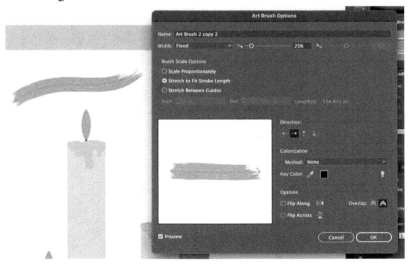

Figure 4.9 – Art Brush Options panel

A beautiful and intuitive **Paintbrush** tool variation is **Bristle Brush**. It allows a wide variety of adjustments to customize the texture of your brush. You can adjust **Size**, **Bristle Length**, **Bristle Density**, **Bristle Thickness**, **Paint Opacity**, and **Stiffness**:

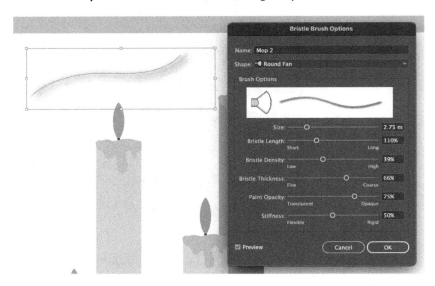

Figure 4.10 – Bristle Brush Options panel

The final brush profile is **Pattern Brush**. This is similar to **Calligraphic Brush** as it repeats a brush tip, but it also allows for alternate patterns for the inside and outside corners. It can also have unique start and end patterns for open strokes:

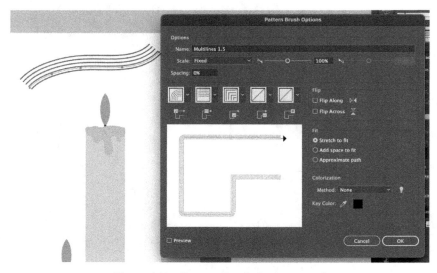

Figure 4.11 – Pattern Brush Options panel

Now that we have reviewed all the varied brush profile choices, you should be able to make a more informed choice as to which one you are using to make your custom brush. Use the following steps to create your own brushes:

4. Create a shape or path that you would like to use as the basis for the brush.
5. Open the **Brushes** panel.
6. Access the **New Brush** command in the upper right or lower right (the + icon) of this panel.
7. Choose the brush profile method you would like to use.

Calligraphic Brush is an exception to this, as you do not have to create a vector shape or path beforehand. Just select the **New Brush** command and adjust the **brush tip** in the options panel that is presented during this process.

Blob Brush tool

In addition to the traditional **Paintbrush** tool, there is also **Blob Brush** (*Shift + B*).

Blob Brush is unique as it does not paint a path, but rather, it makes a vector path around the perimeter of the fill of the brushstroke(s). As you continue to paint, the vector path continues to surround the ever-growing "blob" that you paint. If you are painting with the same color, each additional brush stroke will continue to grow the selection:

Figure 4.12 – Blob Brush use

Using the Pencil and Paintbrush tools will allow you to create hand-crafted objects freely and intuitively, but they may not offer you the control and accuracy that you need for a particular task. Next, we will look at the distinct advantages of both creating with the **Pen** tool and editing at very specific points along any path.

Advanced training on the Pen tool and anchor point editing

So far, we have been focused on creating paths with tools that follow the movement of your mouse, a gesture on the touchpad, or a swipe of a pencil on a tablet.

The **Pen** tool gives additional control by placing **anchor points** and **direction handles** along the path, which you can use to adjust it. The following steps can be used as an introduction if you are somewhat new to this tool:

1. Select **Pen** tool (*P*).
2. Drop an initial anchor point by clicking anywhere on the artboard.
3. Move the tool, and then click again to create a straight line.
4. Hold *Shift* while clicking to constrain the straight line to a multiple of 45˚.
5. Click and drag the **Pen** tool to create the direction handles of the anchor point.
6. click and drag again to choose the angle and scale of the curve segment, and then release.

As noted in steps *3* and *4*, holding down the *Shift* key while drawing allows you to draw straight lines at controlled angles:

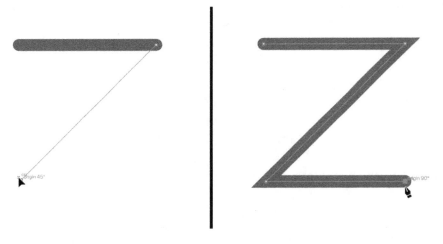

Figure 4.13 – Holding Shift to constrain lines to vertical, horizontal, or 45˚ angles

As noted in steps 5 and 6, you can choose the desired angle and depth of curves using the direction handles:

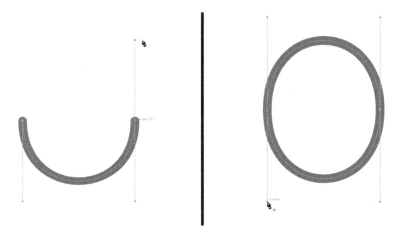

Figure 4.14 – Drawing out directional lines to create angle and depth of curves

In addition to the traditional **Pen** tool, there is also the **Curvature** tool (*Shift + ~*), which allows you to draw with a more intuitive approach. Follow these steps if you would also like an introduction to this tool:

1. Select the **Curvature** tool.

2. Drop an initial anchor point by clicking anywhere on the artboard.

3. Move the tool and then click again to create a straight line.

4. As you move away from the second anchor, a suggested path appears,which allows you to make a better choice as to where the next anchor should be placed for your intended shape.

5. As you continue to drop anchors, the suggested path continues to appear and give you suggestions as to the placement of each additional anchor.

6. Any time you would rather that the anchor creates a corner, just hold *Alt/Option* while placing it.

7. When editing, anchors can also be converted from corner to curve or vice versa by clicking on them while holding *Alt/Option*.

8. You can also edit a shape while you are still on the **Curvature** tool. Just click on any single anchor, and then drag it to its new location.

9. While you are still using the **Curvature** tool, you can also add additional anchors by clicking on the path.

10. You can also convert any of the anchors by simply double-clicking them while still using the **Curvature** tool.

Of all the vector drawing tools within Illustrator, I found the **Curvature** tool was the most difficult to get used to and understand its true benefits. As you continue to use it, the tool's true potential will start to come out. The **Curvature** tool gives you the ability to create smooth curves very quickly through a method of previewing them before committing to them, as shown in this figure:

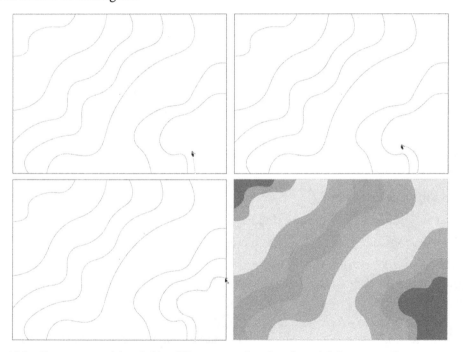

Figure 4.15 – Curvature rool (top left), adding a second anchor (top right), suggested curves as anchors are added (bottom left), and converting corner anchors by Alt-clicking (bottom right). Suggested curves are applied after the final anchor is placed

With the ability to preview the path as it is being developed, you might find there isn't as much need to edit the path after it's been created, but you can still select individual anchors or add additional anchors to the path using the **Curvature** tool. You can also use **Direct Select Tool** to select several anchors at once by holding the *Shift* key while clicking on them. This will be discussed in greater detail in the next subsections, but an example of an adjustment can be seen in *Figure 4.16*:

Figure 4.16 – The Curvature tool aids in creating smooth curves

Figure 4.16 shows the adjustment from *Figure 4.15*, which was created, in part, from the ability to select and edit the anchors of the previously drawn shapes.

The variety of tools that Illustrator offers allows you to create very accurate artwork using methods most appropriate for your specific situation. Having this variety of tools often gets you very close to what you intended, but knowing exactly how to control and manipulate the vector paths down to the anchor level will allow you to master your designs.

Advanced training on the Pen tool

The **Pen** tool is not the easiest tool to get used to, but with these tips, and some practice, you should be able to see its advantages soon. Generally, the **Pen** tool allows you to create either straight path segments or curved path segments, and any combination of the two. The following list will allow you to create open and closed paths with great accuracy:

1.   Select the **Pen** tool.
2.   Drop an initial anchor point by clicking anywhere on the artboard.
3.   Move the tool and then click again to create a straight segment.
4.   Move the tool to a new location and click again to create the next straight segment.
5.   If a curved segment is required along the path, just click and drag to create direction handles for the anchor.
6.   The angle and length at which you pull the handles will impact the curve's size and direction.

7. To continue the curve, move the tool in the desired direction and click to drop a new anchor.

8. If you want to convert the last anchor from corner to curve or vice versa, just hold *Alt/Option* and click on it.

9. If you want to split the direction of an anchor, just hold *Alt/Option* and click and pull on one of the direction handles.

As noted in step 6, you can choose the desired angle and depth of curves using the direction handles. Think of a direction handle as a tool that works to bend and stretch this thin band of wire:

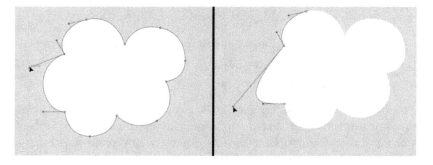

Figure 4.17 – Adjusting curves with direction handles

After becoming more familiar with the **Pen** tool, you will start to get better at deciding what to do to make the intended shape, but even if it all goes well, the shape usually requires some editing. You may consider doing what you've learned and switching to the **Direct Select Tool** (*A*) to edit the path at the individual anchor level. You can most certainly do that, but you should also be aware that it can be done while you are still using the **Pen** or **Curvature tools**. Let's look at that next.

Advanced training on anchor point editing

As I just mentioned, as you get more comfortable with the **Direct Select Tool** and the techniques for editing individual anchors, you should try using keyboard shortcuts to become more efficient with your editing. With keyboard shortcuts, you should be able to edit your anchors as you continue to draw your vector path. I would suggest trying the following shortcuts to speed up your workflow:

- Hover over the last anchor to convert it from corner to curve or vice versa.

- Hold the *Alt/Option* key while clicking and moving the curve anchor handle to create curves on both sides of a corner anchor.

- Hold *Ctrl/Command* to change your cursor to the **Direct Select Tool**. This will allow you to quickly move between selecting the entire shape or just a specific anchor.

- The *Delete/Backspace* key will remove the last anchor point.

Give those shortcuts a try and you might find less and less need for switching continually between the Pen and **Direct Select tools**.

Now that we covered most of the drawing tools within Illustrator, I think it's time to talk about the times when you need to create work with very accurate shapes. The shape tools and **Shaper** tool, along with some advanced modification tools (such as the **Shape Builder** tool and the **Pathfinder** panel) will allow you to create very precise artwork.

Exploring simple shape tools and the Shaper tool to develop more advanced objects

I personally love creating in Illustrator using the simple shapes that the shape tools offer. Let's first look at the simplest drawing methods and then examine some methods for adding and subtracting from combined objects.

Shape tools

As we look at the list of available shape tools, please be aware that the **Flare** tool is also in this same family of tools. We will not be discussing it with the others, as it has a very different purpose and often feels like it should be located with some other tool family. If you are unfamiliar with the tool, it is used to create a simulated lens flare and can be fun to use, but this just isn't the right time to examine it. Instead, we will be sticking to the five main shape tools:

- **Rectangle Tool** (*M*)
- **Rounded Rectangle Tool**
- **Ellipse Tool** (*L*)
- **Polygon Tool**
- **Star Tool**

To select any of these tools you can use their keyboard shortcuts, click and hold on the top tool in the tool family within the toolbar, and then select your choice from the visible icons or use the shape tools tear-off panel. To engage this option, just hold your mouse button down while in the tool family view, move your cursor to the right so it is hovering over the small arrow, and then let go of the mouse button. A new panel of all the shape tools will now be floating in your workspace. If you ever want to close it, just click on the **X** button at the top left of the panel, as shown here:

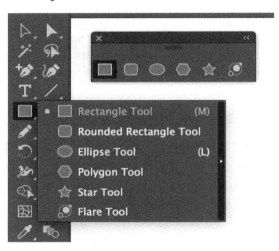

Figure 4.18 – Engaging the Shape Tools Tear-Off panel

You may also use keyboard shortcuts while drawing with these tools to gain additional benefits. Holding the *Shift* key while drawing with the **Rectangle Tool** (*M*) will create a perfect square. Similarly, holding the *Shift* key when using the **Ellipse Tool** (*L*) to create a perfect circle. If you hold the *Shift* key while using either the **Polygon Tool** or **Star Tool**, it will align the object to be level in relation to the base of the document.

The **Rectangle Tool**, **Rounded Rectangle Tool**, and **Ellipse Tool** draw outward from the point of origin (where you began the shape) and continue to grow in the direction you draw to. If you hold the *Alt/Option* key while drawing with these tools, center as the point of origin from which the shape is drawn outward.

Shift and *Alt/Option* modifier keys may be used together to combine their benefits. For example, holding down both of these modifier keys while drawing with the **Rectangle Tool** will result in a perfect square being drawn outward from the point of origin.

If you are drawing with either the **Polygon Tool** or **Star Tool**, you can adjust the number of sides/points by tapping the *up* and *down* arrows of the keyboard while holding the left mouse button down. You will gain more sides/points each time you tap the *up* arrow, and less each time you tap the *down* arrow.

Instead of drawing with your mouse, you can also create any of these shapes by clicking on the artboard after selecting the chosen shape. This will bring up the options panel for the chosen shape. This will give the **Constrain Width and Height Proportions** link for rectangles, rounded rectangles, and ellipses so that you can either make perfect shapes or scale shapes to a common ratio. With the link icon selected, each time you change one of the two dimensions, the other grows or shrinks at the same ratio. In the case of *Figure 4.19*, the rectangles were created using a consistent 1:2 ratio. The first was a 5 by 10 pt rectangle and then the width was increased by 5 each time. To maintain the ratio, Illustrator changed the height by 10 each time, as shown here:

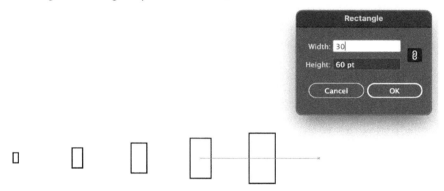

Figure 4.19 – The Constrain Width and Height Proportions link creating a series of shapes

When drawing a polygon or star, clicking on the artboard will again bring up the options panel for them, but will allow for the number of sides/points and the shape's radius. In the case of a star, **Radius 1** is the outer radius, and **Radius 2** is the inner radius:

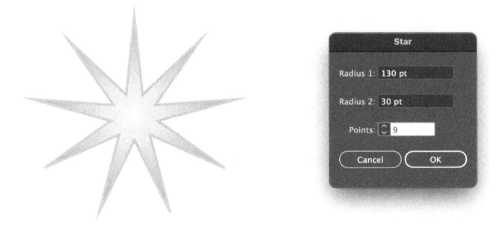

Figure 4.20 – Inner and outer radius using Star Tool's options panel

If you are drawing with the **Rounded Rectangle Tool**, you can use the keyboard's directional arrows to adjust the radius of the corners. While holding down the *up* arrow, you will slowly increase the radius of the rectangle, and holding down the *down* arrow will slowly decrease the radius. Tapping the *right* arrow will result in a maximum radius while tapping on the *left* arrow will result in a minimum (no) radius. I use the *left* and *right* keys to get closer to the radius and then apply either the *up* or *down* key to move from either the maximum or minimum radius. Keep in mind that these shortcuts will only work while continuing to hold down the mouse button. After releasing the mouse button, you can still adjust the radius of all the corners by using the **Selection** tool to select the shape and then click on and move one of the **Live Corner widgets**. If you would like to adjust one or more (but not all) of the corners, use the **Direct Selection** tool – click on any corner anchor, hold *Shift* while selecting the second and/or third corner anchor, and then click on and move any of the **Live Corner widgets**.

In addition to the shape tools, Illustrator also offers several unique tools that focus on lines instead of shapes. The **Line Segment** tool (\) allows you to click and drag with the left mouse button held to draw a straight-line segment at any angle. In addition, if you hold the *Shift* key while drawing out the line, it can travel horizontally, vertically, or at a 45˚ angle, depending on the direction your mouse travels. Holding down the *Alt/Option* key, in this case, will allow the line segment to grow outward evenly from the point of origin.

Just like the previous shape tools, you can also select The **Line Segment** tool and then click on the artboard to bring up the options panel. The **Length** and **Angle values** in the options panel will always match the previous line segment drawn, so it is an excellent way to place duplicate line segments right where you want them. Just click the desired location, and hit *Enter/Return* to accept the dimension and angle:

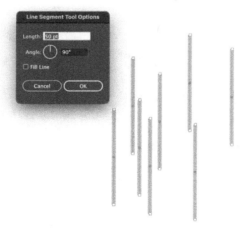

Figure 4.21 – Using the Line Segment tool Options panel

Within the same family of tools is the **Arc** tool. This tool allows you to make custom, smooth arcs. With some adjustments in the tool's options panel, you can also make closed shapes that can be filled or unfilled. Clicking anywhere on the artboard while using the tool will present you with the options panel. See *Figure 4.22* for the **Type** choices that **Arc** tool offers you, as well as **Length X-Axis** and **Length Y-Axis**, **Slope**, and **Base Along X-Axis** or **Y-Axis**:

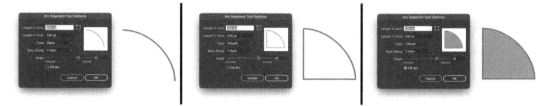

Figure 4.22 – Using the Arc Segment Tool Options panel to create an arc,
an unfilled shape, and a filled shape

In addition to simply drawing out arcs with this tool or using the tool's options panel, by clicking on the artboard or double-clicking on the tool's icon in the toolbar, you can also adjust the tool by utilizing the same familiar keyboard shortcuts you might use for other drawing tools. If you hold *Shift*, it will constrain to a 45° angle between the two anchors making up the arc. Holding *Ctrl/Command* will allow you to draw the arc outward from the point of origin. Tapping or holding the *up* arrow will increase the slope while tapping or holding the *down* arrow will decrease the slope. Remember that this will only work while you continue to hold down the mouse button.

In addition to all these techniques for path tools and the adjustment of the paths they draw, the one that is by far the most unique is the *tilde* (~) key; although it has limited usage, I believe you will find it interesting if you are not yet familiar with its results. The benefits of the *tilde* (~) key will be discussed in *Chapter 5*, *Editing and Transforming Objects*.

Shaper tool

The **Shaper** tool (*Shift + N*) will attempt to understand your gestures. If you draw a path with a circular shape, a perfect circle magically appears. This will also work for ovals, triangles, squares, rectangles, and hexagons:

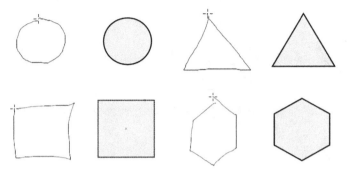

Figure 4.23 – Shaper tool converts a free-hand gesture into a circle, ellipse, rectangle, triangle, or another polygon

This tool also allows you to use free-hand gestures to create additional shape constructions. Be sure to start with a collection of overlapping shapes, which can be created using the **Shaper** tool or any other drawing tool. Then, select the **Shaper** tool and scribble within a shape or overlapping area to subtract the area (any group of overlapping shapes will automatically be put into a Shaper Group and highlighted while using the **Shaper** tool). For example, I have scribbled over the overlapping area of the red rounded rectangle and the green circle in *Figure 4.24*. This figure shows the resulting area being punched out of the Shaper Group:

Figure 4.24 – Subtraction of overlap using a gesture

In addition to using a mouse gesture to create this scribble, if available, you can also use either a stylus and a tablet or a touch screen to complete the gesture.

In *Figure 4.25*, applying a scribble gesture over the area of the blue triangle not overlapping the green circle has resulted in the removal of the non-overlapped portion of the triangle:

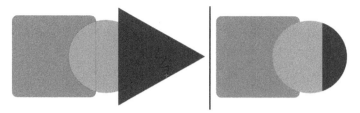

Figure 4.25 – Subtraction outside overlap using a gesture

If you were to make a similar gesture on the portion of the rounded red rectangle that was not overlapping the green circle, the same action would result, and the green color would remain on the overlapped area.

You can also merge shapes using this method of scribble gestures. As you can see in *Figure 4.26*, I have scribbled across all the shapes to merge them into one. The merged Shaper Group will be colored with the color of the shape at the point of origin of the gesture. In this case, it is red, as I began on the red rounded rectangle and ended on the blue triangle, as shown here:

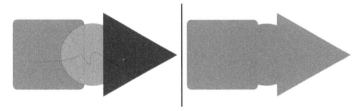

Figure 4.26 – Addition of overlap using gesture

If you are starting with a shape behind the others, then scribble from the non-overlapped area to the overlapped area to merge them. If you are trying to merge to the top shape, then start on the overlapped area and scribble back to the underlying shape. This will allow you to retain the top shape's color upon merging the two.

After using the **Shaper** tool in this way, it will assign these shapes to a Shaper Group and allow them to remain interactive. To interact with any of these new groups, select the **Shaper** tool and then select one of these groups. It will become selected and display an *arrow* widget. If the arrow is pointing down, it indicates that you can select the entire group. Clicking the objects in the group will make it active and it can then be edited as one object or can continue to be edited with the scribble gestures. The top two screenshots in *Figure 4.27* illustrate how the group looks after being selected:

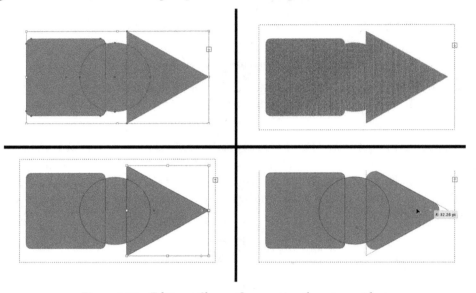

Figure 4.27 – Editing a Shaper Group using the arrow widget

The bottom two screenshots in *Figure 4.27* illustrate that you can also engage any specific shape that originally made up this now merged object. To do this, you will need to click the *arrow* widget and it will then change to an arrow pointing upward. Now, you can click on any individual shape and make the desired adjustments. In the case of *Figure 4.27*, I adjusted the radius of the triangular portion of the Shaper Group.

Next, let's look at a tool that can be used very similarly, but that requires you to simply draw a line, or at times, requires just a simple click of the mouse.

Combining shapes with the Shape Builder tool

If you aren't familiar with this technique, I am confident that you will find it useful and quite enjoyable to do. It has many similarities to the **Shaper** tool, but I find it simpler and more intuitive, as it not only allows you to combine and remove areas by using overlapping shapes, but it also considers every overlapping area as a potential new shape. For example, you could join two overlapping areas without joining the original shape. Just like the **Shaper** tool, it can be used with overlapping shapes that were created with any of the drawing tools:

1. Select all objects you intend to use with the **Shape Builder** tool.

2. Select the **Shape Builder** tool (*Shift + M*).

3. Draw through the shapes you would like to unite. They will join to create a new shape.

4. Hold down the *Alt/Option* key and click or draw through shapes you would like to remove. They will be subtracted from the rest of the selection.

5. Hold down the *Shift* key in addition to the *Alt/Option* key while drawing through the shapes to allow them to be selected within a selection rectangle. Also, you can hold *Shift + Alt/Option* to subtract within a selection rectangle.

Figure 4.28 illustrates that drawing a long meandering line (left) can result in several shapes being merged into one (right):

Figure 4.28 – Drawing through multiple shapes to unite

I especially like this technique for constructing shapes, but remember to try out each method and consider how they each have their benefits and drawbacks. It really depends on the use case. There are times when these techniques are not enough and you need something with more options. That's where the next tool comes in.

The Pathfinder panel for heavy lifting

If the situation does not allow for simple addition or subtraction of combined elements, then it might be time to use the advanced properties of **Pathfinder**. **Pathfinder** was in Illustrator long before the **Shape Builder** tool, but I would now recommend the **Shape Builder** tool as a primary option and **Pathfinder** as a secondary option when needed. It allows for more unique options for blending multiple objects.

As *Figure 4.29* shows, the **Pathfinder** panel is divided into two areas: **Shape Modes** and **Pathfinders**:

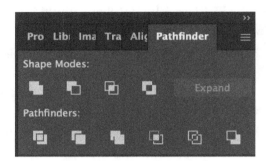

Figure 4.29 – The Pathfinder panel

To illustrate this tool, we will use a sample image made up of several shapes and show the results of each method. This artwork will be used to run through all **Shape Modes** and **Pathfinders**, and then I will share the results from the two categories:

Figure 4.30 – Original

Figure 4.31 shows the results after applying each of the shape modes:

Figure 4.31 – Shape Modes

The original, as you can see in *Figure 4.30*, is made up of two grouped sets of shapes, so it results in an appearance you would expect each time. In *Figure 4.31*, this is how each of the shape modes works:

- **UNITE** puts both parts together and uses the attributes of the top object.
- **MINUS FRONT** removes the overlapping area that was taken up by the top group of shapes.
- **INTERSECT** results in the areas that both groups have in common.
- **EXCLUDE** results in the shapes that are different in both groups.

Figure 4.32 shows the results after applying each of the Pathfinders:

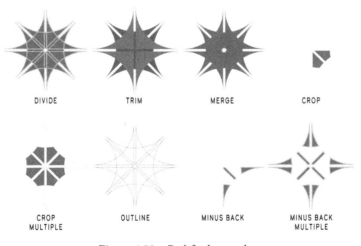

Figure 4.32 – Pathfinder modes

Let's discuss each of the Pathfinders in *Figure 4.32*:

- **DIVIDE** splits all the enclosed vector paths into individual objects.

  (To edit the newly created objects, you will need to ungroup the result.)

- **TRIM** removes any strokes but doesn't merge fills.

- **MERGE** removes any strokes, as well as merges any of the same colors that are overlapping.

- **CROP** removes any strokes and then deletes all but the top-most object (note that multiple copies of the original shapes were used to create the **CROP MULTIPLE** result).

- **OUTLINE** divides the objects into their line segments (useful for creating a trap for overprinting).

- **MINUS BACK** removes the overlapping area that was taken up by the bottom group of shapes (again, note that multiple copies of the original shapes were used to create the **MINUS BACK MULTIPLE** result).

We have now examined several methods to draw, edit, and combine objects. Next, let's look at a method that allows for even greater flexibility for later editing when combining objects.

Compound paths

Compound paths are when two or more paths are combined so that holes appear where paths overlap. You can create this by selecting both objects that have a common overlap, right-clicking, and choosing **Make Compound Path** from the contextual menu. You can also choose to go to **Object** > **Compound Path** > **Make** *(Ctrl/Command + 8)* from the menu bar to create the compound path. All objects in the compound path will take on the attributes of the backmost object in the stacking order.

As you can see in the demonstration of the sun image in *Figure 4.33*, the holes created from a compound path allow you to show underlying objects:

Figure 4.33 – (L to R) Compound path, simple path, and compound path over simple path

An additional benefit to this method is the ability to individually edit the members of the compound path.

Compound shapes

This last example shows the benefits of the compound shape. It allows you to hold multiple shapes together as one new shape but has the benefit of allowing further editing of the individual components within the compound shape. You can create a compound shape by holding down the *Alt/Option* key while selecting any of the shape modes in the **Pathfinder** tool. See *Figure 4.34* for a detailed example:

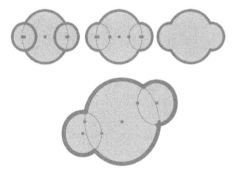

Figure 4.34 – Top (L to R) – Multiple shapes, compound shape, and compound shape expanded; Bottom – Adjusted compound shape after being expanded

Double-click any individual shape within a compound shape to enter **Isolation Mode** and edit it individually. When you have completed the edit, you will need to click out of **Isolation Mode** by clicking the left-pointing arrow in the **Isolation Mode** bar (see *Figure 4.33*):

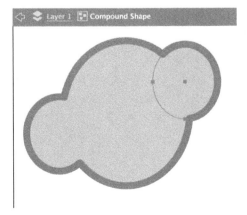

Figure 4.35 – Compound shape being edited in Isolation Mode

You might note that a difference between a compound path and a compound shape is that the latter will not create a hole with any of the objects included in the newly joined creation.

Summary

After having completed this chapter, you now know how to use a variety of drawing tools that come with Adobe Illustrator. You also now know how to build more complex illustrations using these tools individually or in combination.

We have reviewed the attributes of each tool and several techniques for utilizing them to a high level of their capabilities.

All the information within this chapter was intended to challenge you to venture out with your creativity and find new ways to transfer it to the artboard using this variety of tools to deliver your concepts.

In the next chapter, we will be discussing several techniques for editing and transforming your newly drawn objects. You will learn that you have complete control of the shape and form of your vectors and can reshape their paths at any time. We will also explore some of the clever additional methods of manipulating their paths, through options and effects.

5
Editing and Transforming Objects

Now that you have verified or increased your understanding of the main methods available for creating in Illustrator, let's go a little deeper to increase your skills and efficiency in both creating and editing vector paths and objects. We will look at methods for editing paths *after* they have been created, as well as how you might edit them *while* they are being created. We will also explore methods for transforming your paths and several of the exciting ways you can apply effects to them.

In this chapter, we are going to look specifically at three methods of editing and transforming your work. We will begin by editing parts of an object and then continue by transforming and enhancing objects.

To accomplish this, the chapter will be divided into the following main topics:

- Anchor-level editing
- Transformations and their options panels
- The Effects panel and its options

Technical requirements

To complete this chapter, you will need the following:

- Adobe Illustrator 2022 (version 26.0 or above).
- High-quality internet access may be required for some situations.

Anchor-level editing

Anchor-level editing gives you complete control of any path. Being able to select specific anchors could allow you to create a raindrop from a circle, for example. Selecting just the top anchor of the circle and then moving it upward could allow you to have a very nice droplet shape.

It is very important that you learn the benefits of editing anchors either individually or in selected groupings. The **Selection** tool (*V*) allows you to pick an entire object, the **Direct Selection** tool (*A*) allows you to select any individual anchor so that you can make further refinements to it. In addition, you can add additional anchors by holding down the *Shift* key while clicking them.

Looking at the following figure, you can see that the first adjustment was made by selecting one anchor with the **Direct Selection** tool and lowering it, the second adjustment was made by holding *Shift*, selecting two anchors, and then lowering them, and the final adjustment was made by holding *Shift*, selecting three anchors, and then moving them to the right:

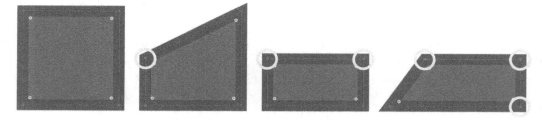

Figure 5.1 – Anchor edits (L to R) – original rectangle, one anchor selected and adjusted, two anchors selected and adjusted, and three anchors selected and adjusted

A great shortcut that can be used when either **Selection** tool or the **Direct Selection** tool is selected, is to hold down the *Ctrl/Command* key to quickly flip to the opposite tool. If you are currently using **Selection** tool, holding the *Ctrl/Command* key will change it to the **Direct Selection** tool, and vice versa. When you release the *Ctrl/Command* key, you will again be able to use the tool you currently have selected.

For example, if I am currently using the **Selection tool** to move an object and then want to quickly select just one anchor and adjust it, I can switch to the **Direct Selection** tool by holding down the *Ctrl/Command* key, make the adjustment, and then release the *Ctrl/Command* key to continue with the **Selection tool**.

In addition to being able to select an individual or group of anchors and changing their position, you will also be able to edit the anchor's corner radius and direction handles:

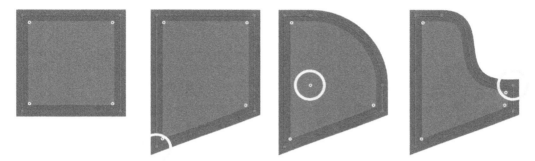

Figure 5.2 – Direct Selection edit options (L to R) – original rectangle, lower-left anchor selected and then moved, upper-right anchor selected and then corner radius adjusted, and curve anchor selected and then angle adjusted using the direction handle

The ability to select any anchor or (group of anchors) allows for further editing of any created path. This alleviates a lot of stress, as you can now create knowing that everything can be later adjusted. You want to try and design with the fewest anchors possible, as this will assist you in keeping your artwork smooth. In addition, it will allow your system to display and print them faster. An additional benefit of drawing with a low number of anchors is that you will have fewer anchors to edit at the anchor level.

Follow along with this short project to practice editing at the anchor level as we create a candle out of just three simple shapes:

1. Draw a rectangle using **Rectangle** tool (see *Figure 5.3, top left*). This is intended to create the body of the candle, so choose the size and color attributes that you feel represent the object well.

2. Choose the **Direct Selection** tool and then select the top two anchors by clicking them while holding *Shift*. You could also select both anchors by dragging the **Direct Selection** tool around the two anchors. This will only work if there isn't anything else around them that could also be selected.

3. Grab the **Live Corners** widget (the small white circle now visible near each anchor) for either of the anchors and pull toward the center until the two Live Corners widgets meet (see *Figure 5.3, top center*).

4. Again, choose the **Direct Selection** tool and then select the top of the arch that was created in step *3*. It is difficult to see, but there are two anchors there, so you will need to select them just as you selected the two corners in step *2*. Select the two anchors by clicking them while holding *Shift* (clicking to the left and right of the anchor area will allow you to get them both). You could also select both anchors by dragging the **Direct Selection** tool around the two anchors. This will only work if there isn't anything else around them that could also be selected (see *Figure 5.3, top right*).

5. Using the **Direct Selection** tool, pull the selection you made in step *4* downward while holding *Shift* to constrain movement vertically (see *Figure 5.3, bottom left*). This will allow you to create the dip in the side of the melting candle (see *Figure 5.3, bottom center*). As an alternative, you could also omit holding down *Shift*, and then the dip will become asymmetrical and give the melting candle a more random feeling.

6. Draw a new rectangle shape to create the wick of the candle and choose the size and color attributes that you feel represent the object well (see *Figure 5.3, bottom right*). Be sure to adjust its layer order to be arranged below the candle body layer.

Here are the visual examples for each step in this process:

Figure 5.3 – Anchor-level editing to create a candle

Now that we have successfully designed a body for the candle, we will create the flame with only one more shape and some anchor-level editing:

1. Before beginning the creation of the flame, be sure that the wick shape is centered with the candle body (see *Figure 5.4, top left*).

2. Draw out a long ellipse to represent the flame and fill it with a warm gradient (see *Figure 5.4, top center*). Using the **Gradient** tool, adjust the gradient to your liking and move the brightest area over the end of the wick by selecting the small black dot at the end of the widget and dragging it to the new location.

3. Using the **Anchor Point** tool (*Shift + C*), click the top anchor to convert it to a corner point (see *Figure 5.4, top right*).

4. Select the top anchor (which you just converted in step *3*) with **Direct Select Tool** and then move it to the right. I chose to use the right navigational arrow on the keyboard while also holding down the *Shift* key to move it 10 pixels each time. You may choose any distance that looks good to you, but I chose to move it five times.

5. While this top anchor is still selected, choose the **Anchor Point** tool, and click and drag the top directional point of the anchor on the right of the ellipse (see *Figure 5.4, bottom left*). This will convert the anchor to a corner point.

6. Use **Direct Select Tool** to select the anchor that was converted in step *5* and then pull the Live Corners widget (see *Figure 5.4, bottom center*) towards the center until it creates an appropriate curve for the appearance of the flame (see *Figure 5.4, bottom right*).

Here are the visual examples for each step in this process:

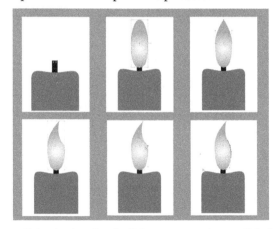

Figure 5.4 – Anchor-level editing to create the candle's flame

Completing this short project has allowed you to practice editing at the anchor level while creating an object with just three simple shapes. Keep experimenting with what you can create by using slight adjustments to your basic shapes, and you will soon discover that many more complex ideas can be generated.

Another way we can quickly make adjustments to either selected anchors or entire objects is through transformations. In the next section, we will explore their uses and benefits for creating more advanced illustrations.

Transformations and their options panels

Transformation methods help you to adjust elements within your design and increase your ability to maintain symmetry and overall balance. We have five types of transformations at our disposal. To access these options, you can choose **Object** > **Transform** or right-click on any selected object. Let's take a closer look at each of the options:

- **Move** (*Shift + Ctrl/Command + M*): This transformation allows you to enter specific measurements of movement and the directional angle. It can be used to move the object, the pattern fill, or both (see *Figure 5.5*):

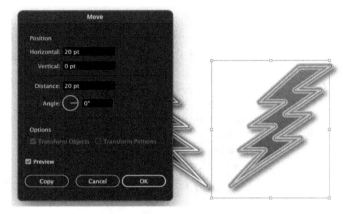

Figure 5.5 – Transform > Move > Horizontal

- **Rotate**: With this, you can rotate the object, fill the pattern, or both to a specific angle (see *Figure 5.6*):

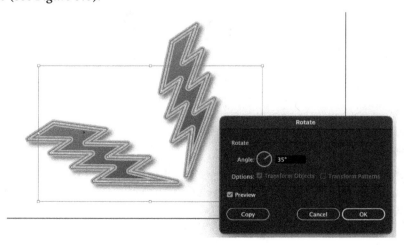

Figure 5.6 – Transform > Rotate > Angle

- **Reflect**: I find this to be one of my most often used options of the group, as it allows you to quickly flip an object from left to right (and vice versa) or top to bottom (and again, vice versa). It can also flip or "mirror" the object using any angle, and can affect the object, pattern, or both (see *Figure 5.7* for an example of this):

Figure 5.7 – Transform > Reflect > Vertical

- **Scale**: This has more options than you might expect. Of course, it allows for a basic scale adjustment, either up or down, but it also offers several additional options (see *Figure 5.8*):

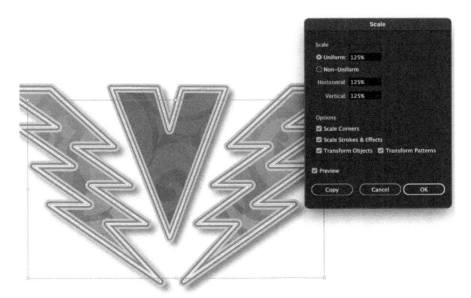

Figure 5.8 – Scale transformation panel

- **Shear**: This option allows your selected anchors, object(s), or group to slide at any angle you choose. Like the other options, it can affect the object, pattern, or both. I encourage you to use the **Preview** checkbox on any of these tools. It is a great way to see the results without making a commitment (see *Figure 5.9*):

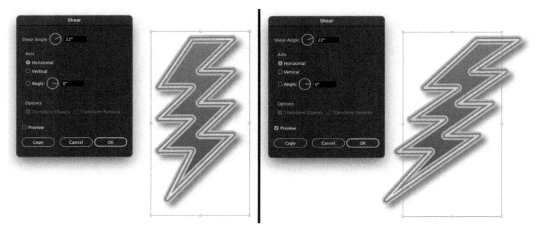

Figure 5.9 – Utilizing Preview in the Shear option panel

- **Transform Each** (*Alt/Option + Shift + Ctrl/Command + D*): This can be your "one-stop-shop" version of the transformation tools. It allows you to adjust the scale, position, angle, and reflection, all in one panel. In addition, it allows you to make the adjustments to more than one of them at one time (see *Figure 5.10* for an example of this):

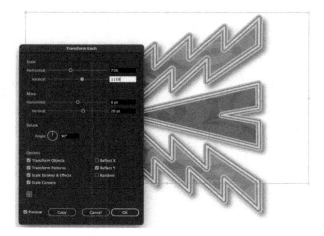

Figure 5.10 – Multiple adjustments from within the Transform Each panel

- **Transform Again** (*Ctrl/Command* + *D*): This last option is a great way to duplicate the last transformation you performed, and I would highly recommend you learn the shortcut so you can perform an action repeatedly.

As an example of these transformation methods being put to action, consider the clouds in *Figure 5.11*. Pay close attention to the limited number of base shapes, and how they were copied and varied using the previously mentioned transformation methods:

Figure 5.11 – Create variations of original objects using the Copy command

Transform Again is a great way to duplicate the last transformation, but remember that each of the tools in the transformation category offers a **Copy** button, which allows you to create variations of the original.

We have now reviewed several ways by which you can edit elements in your illustration, yet there are still many additional options available in the **Effect** panel. We are going to look at these next and highlight several of them in a step-by-step project.

The Effect panel and its options

If you are already familiar with **Adobe Photoshop**, then you will likely be familiar with the **Filter** menu in that application. If you are not, I would describe this menu as a list of options that will render visual adjustments to the selected item(s). Illustrator has something that is quite comparable to what you would experience in Photoshop. In fact, one of the biggest differences you may notice is the actual name. While the list of available options in this menu are filters in Photoshop, they are considered effects in Illustrator and are divided based on whether they process the object using vectors or pixels. Any third-party software you have installed will be in a third category. You will find all these categories in the **Effect** menu, located in the top menu:

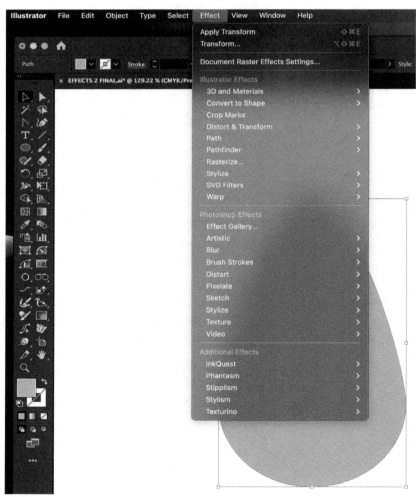

Figure 5.12 – The effects categories within the Effect menu

In addition to dividing the effects into appropriate categories, the **Effect** menu also presents two useful shortcuts to you, which are located at the top of this menu:

- **Transform…** (*Alt/Option + Shift + Ctrl/Command + E*) allows you to open the last used effect. You would then be able to make changes with the **Effect** panel before accepting.

- **Apply Transform** (*Shift + Ctrl/Command + E*) allows you to run the last used effect. This method will create the exact same result that the effect created previously and will not present an input panel for making any changes to the effect.

As you go down the list from these two shortcuts, you will find the effects divided into these categories:

- The first category is **Illustrator Effects**. This first category will allow you to adjust any object, group, or layer, and change its characteristics.

- The second category is **Photoshop Effects**, and they are raster effects. Using these will result in raster-based content. These effects can be applied to either vector or bitmap objects.

- The final category is **Additional Effects**, and it is where you will find any additional effects you have installed from third-party software vendors.

You will also notice the **Document Raster Effects Settings (DRES)** selection in the **Effect** menu. This option allows you to make quality adjustments for your raster effects based on your output needs.

To better explain the benefits of utilizing effects, I want to help you create a quick illustration. Create a new file in Illustrator and try to follow along to discover several methods of using effects. We will be using a **Letter** sized document. If you are opening Illustrator, select **New file**, or you can select the **Letter** preset from the **Presets** panel of the welcome screen (see *Figure 5.13*):

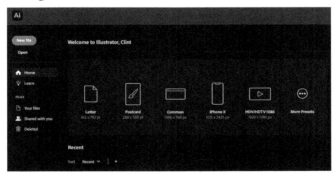

Figure 5.13 – The Presets panel located within the welcome screen

Alternatively, if you already have Illustrator open, you can choose **File** > **New...** (*Ctrl/Command + N*) from the top menu.

Now that you have a newly created artboard, we are going to create a quick illustration of a stylized flower that takes full advantage of the effects:

1. Select the **Ellipse** tool and create a circle in the center of the document. To accomplish that, be sure to hold down the *Alt/Option* key to draw out from the center, as well as the *Shift* key to create a perfect circle. Assign a dark green **Fill** attribute and no **Stroke** attribute.

2. Choose **Effect** > **Distort & Transform** > **Pucker & Bloat...** and set the slider to **-50** (**Pucker**) to create a four-point shape. Deselect this object before moving to the next step. To do so, go to **Select** > **Deselect** (*Shift + Ctrl/Command+A*) in the top menu or hold the *Ctrl* key while clicking outside the selection (see *Figure 5.14*):

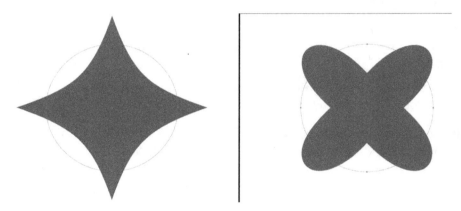

Figure 5.14 – The Pucker & Bloat effect using default circle anchors

3. From the **Appearance** panel, select the **Pucker & Bloat...** content and delete it. This will allow you to create the next shape without maintaining the previous effects applied.

4. Using the **Ellipse** tool again, create another circle that is just slightly smaller than the previous one. For this shape, we will be adding additional anchors before applying an effect. Assign a new color to this section.

5. Go to **Object** > **Path** > **Add Anchor Points**. This command will add anchor points between current anchor points, but if you would like to see them, you will need to select the object with **Direct Select Tool**. Repeat this step one more time for your circle to result in having 16 anchor points.

6. Return to **Effect** > **Distort & Transform** > **Pucker & Bloat...** and set the slider to **25** (**Bloat**) to create 16 petals. Again, be sure to deselect this object before moving to the next step (see *Figure 5.15*):

Figure 5.15 – The Pucker & Bloat effect after adding anchors

7. Once again, from the **Appearance** panel, select the **Pucker & Bloat** content and delete it.

8. Create one more circle in the center of the illustration and assign a new color to this selection.

9. Choose **Effect** > **Distort & Transform** > **Zig Zag...** and set **Size** to **4 pt Absolute**, **Ridges per segment** to **2**, and **Points** to **Smooth**. Note that this effect is not dependent on the number of anchors a path has (see *Figure 5.16*):

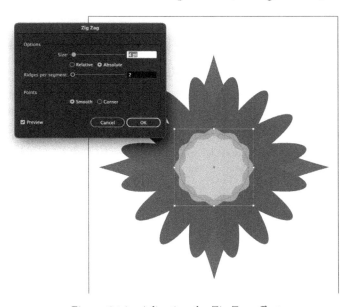

Figure 5.16 – Adjusting the Zig Zag effect

10. While this shape is still selected, add a stroke color, and increase **Stroke** to **16 pt**.

11. Choose **Effect** > **Path** > **Offset Path...** and extend this path with an offset great enough to fill the center of the object (see *Figure 5.17*):

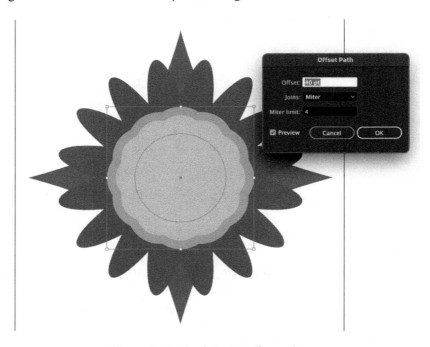

Figure 5.17 – Applying an offset path

12. Select all three Offset Path effect shapes. You can do that by either drawing a large selection rectangle over them or by clicking on each of them while holding down the *Shift* key. Alternatively, you can also target multiple objects by clicking on each of their selection boxes in the **Layers** panel while holding the *Shift* key.

13. Choose **Effect** > **Stylize** > **Drop Shadow** and adjust **Opacity** to **30%**. Note that all the effects to this point have been located under the **Illustrator Effects** section (see *Figure 5.18*):

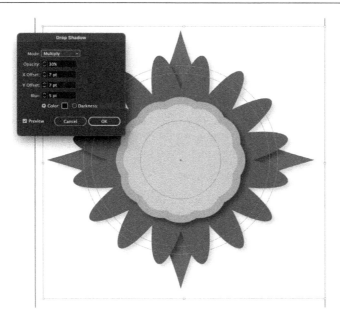

Figure 5.18 – Applying a Drop Shadow effect to multiple objects

14. Create one more circle in the center of the illustration and assign a new color to this selection. Assign no color to the **Stroke** attribute.

15. Choose **Effect** > **Stylize** > **Scribble…** and make adjustments within the dialog box according to your taste or use my settings (see *Figure 5.19*):

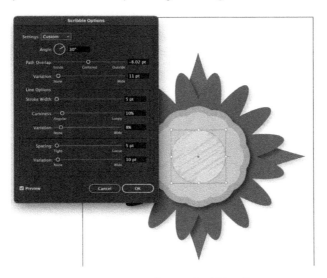

Figure 5.19 – Applying a Scribble effect

16. Next, select the **Line Segment** tool and click and pull down from the center anchor of the previous shape (be sure nothing is currently selected) while holding *Shift* to draw a vertical segment. Stop and release the mouse when you have landed on the bottom boundary line of the artboard. Add a green color to the **Stroke** attribute and increase its stroke width to **24 pt**.

17. While the path is still selected, choose **Object** > **Arrange** > **Send to Back** (*Shift + Ctrl/Command + []*) to arrange it behind all the previous shapes.

18. Choose **Object** > **Expand** from the top menu and leave the **Fill** box checked so that the resulting shape is a green rectangle.

19. Choose **Effect** > **Warp** > **Arch** and select the **Vertical** button and a bend of **15%** (see *Figure 5.20*):

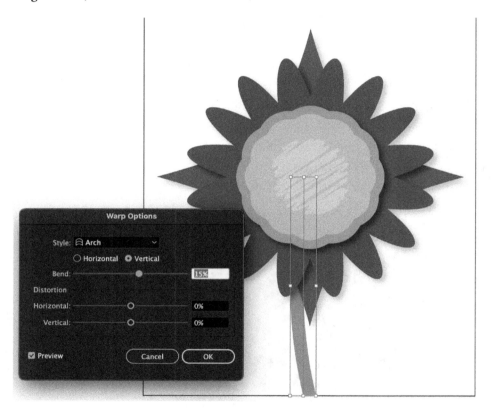

Figure 5.20 – Applying a Warp effect

20. Next, draw an oval using the **Ellipse** tool, and then, using the **Selection** tool and holding down the *Alt/Option* key, move a duplicate that overlaps the original. Add the same green color to the fill that was used for the line in step *16*.

21. After selecting both, choose **Option** > **Group** (*Ctrl/Command* + *G*) and then choose **Effect** > **Pathfinder** > **Intersect** to create the appropriate shape. Right-click on the shape and choose **Transform** > **Rotate** and set the angle to **-15** (see *Figure 5.21*):

Figure 5.21 – Pathfinder intersect

22. To complete this image, use **Rectangle Tool** to create a background that completely covers the artboard. Select a gradient to fill this shape.

23. While it is still selected, create a copy by using **Copy** (*Ctrl/Command* + *C*) and **Paste in Front** (*Ctrl/Command* + *F*).

24. Now, we are going to apply one of the Photoshop effects to add some texture to the background. Choose **Effect** > **Pixelate** > **Color Halftone** and apply **24 Pixels** to **Max. Radius**. Finally, choose **Window** > **Transparency** and set **Blending Mode** to **Multiply** (see *Figure 5.22*):

Figure 5.22 – Color Halftone effect used for background

25. As a final enhancement to this exercise, let's add another level to the petals. Select the current layer and make a duplicate using the same technique we employed for copying the background rectangle. Change the color of **Fill** and then choose **Effect** > **Distort &Transform** > **Twist**. Set **Angle** to **30** (see finished illustration in *Figure 5.23*):

Figure 5.23 – Final creation utilizing effects

In this tutorial, I highlighted several effects that I enjoy using, but I would highly recommend you continue to experiment with the remainder of the available effects so that you can apply them when the appropriate need arises.

Summary

After completing this chapter, you now know how to edit and enhance elements of your work in several ways. We have reviewed how to edit an object at the anchor level. We have also reviewed how to make adjustments using transformations and effects.

All the information within this chapter was intended to increase your skills and knowledge towards adjusting and enhancing your work.

In the next chapter, we will be discussing several additional techniques for customizing objects in your illustration. You will learn several advanced methods of enhancing and adjusting the **Fill** and **Stroke** attributes of any object(s). You will also explore some of the options available for working with color in your illustrations.

6
Advanced Attribute Design

Now that we have discussed both drawing objects (in *Chapter 4*, *Drawing with Pencil, Paintbrush, Pen, and Shape Tools*) and editing objects (in *Chapter 5*, *Editing and Transforming Objects*), it is now time for us to look at some advanced methods for adding attributes to these objects. We will be looking at several techniques for customizing the look of these objects using variations of the two basic attributes that Adobe Illustrator offers for each vector path: fills and strokes. We will explore the **Fill** options and greater color control, as well as the **Stroke** options and a few more advanced time-saving, attribute-adjusting methods.

In this chapter, we will discuss the following topics:

- Fill options and color control
- Multiple fills and strokes through the **Appearance** panel
- Variable width profiles and vector pack brushes for strokes
- Dynamic tools for greater options

Technical requirements

To complete this chapter, you will need the following:

- Adobe Illustrator 2022 (version 26.0 or above).
- High-quality internet access may be required for some situations.
- Although not required, a good-quality drawing tablet or touchscreen is recommended.

Fill options and color control

To customize your work, you have three options to select from, for both fills and strokes. **Fill** and/or **Stroke** could be given a solid color, a gradient, or a pattern. This was first pointed out in *Chapter 1*, *Building a Foundation beyond the Basics*, but we are now going to look at a few advanced methods for working with these attributes.

There are several methods for selecting the fill of your newly created shapes, and we'll review the basics before discussing the more advanced methods that are available to you. After selecting any object using the **Selection** tool (*V*), you can use the **Swatches** panel to select one of the three **Fill** types (**Solid**, **Gradient**, or **Pattern**) to apply to either **Fill** or **Stroke**. In *Figure 6.1*, you can see the **Swatches** panel being opened after selecting the **Fill** box within the Options bar (the Options bar is available within all workspaces except **Essentials**):

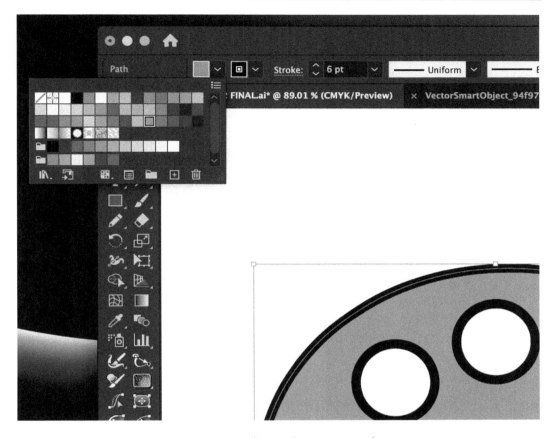

Figure 6.1 – Applying colors using swatches

From the **Swatches** panel, you can select from the default options or add additional swatches or an entire swatch library. You can do this by clicking on the **Swatch Libraries** menu (lower-left icon in the **Swatches** panel) and then selecting from the available libraries.

After selecting a fill, it can be further edited using an appropriate method. For solid fills, you can adjust them using **Color Picker**; for gradient fills, you can use the **Gradient** panel, and for a pattern fill, you can adjust its colors using the **Recolor Artwork** button in the **Quick Actions** panel or the Options bar. If you would rather customize your own color instead of using a swatch, you can hold down the *Shift* key while clicking on the **Fill** box in the Options bar.

You can use the **Options** panel (top-right corner icon of this color selection panel) to select which color space you would like to use. The top choices are **Grayscale**, **RGB** (red, green, and blue), **HSB** (hue, saturation, and brightness), **CMYK** (cyan, magenta, yellow, and black), and **Web Safe RGB** (216 web colors common to all browsers). In addition, there are **Invert** and **Complement**, if you would like to flip the current selection to an opposing color. Follow these steps to mix the exact color you are looking to create:

1. Hold the *Shift* key while clicking on the **Fill** box in the Options bar.
2. Move the cursor over the chosen color spectrum (such as **CMYK Spectrum** while in **CMYK**, **RGB Spectrum** while in **RGB**, and so on).
3. The cursor should now have the appearance of an eye dropper. Select your chosen color from the spectrum by clicking within the color selection box.
4. Modify the color by manipulating the percentile sliders. You can fine-tune the color using this method, as shown here:

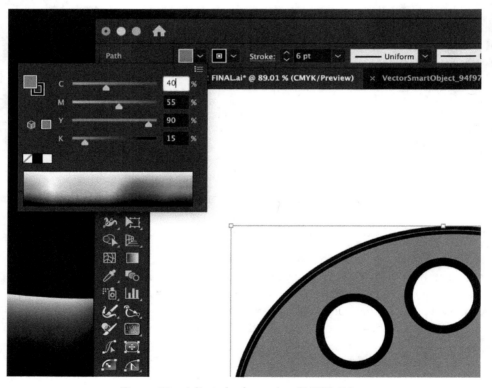

Figure 6.2 – Adjusted colors using CMYK sliders

Figure 6.2 also illustrates that you can directly affect the sliders by inserting the cursor in the percentile box and typing in any desired amount.

Yet another way you can communicate your intended color choice to Illustrator is to enter the hexadecimal or **hex color code** directly into the input box located within the **Color Picker** panel. The input box is represented by the # symbol. See *Figure 6.3* to see its exact location:

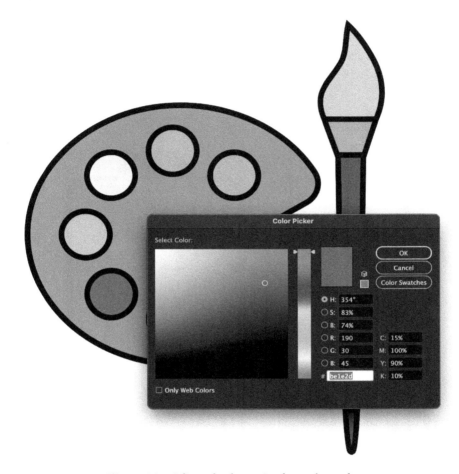

Figure 6.3 – Adjusted colors using hex color codes

In addition to having a variety of ways to choose the **Fill** attribute for any object, one that can save you quite a bit of time (and is a lot of fun) is **Live Paint Bucket** (*K*). Using this tool takes a bit of work at first, but when you are familiar with the process, it proves to be an excellent option for filling lots of shapes in a short amount of time. Follow these steps to use **Live Paint Bucket** on any illustration you create:

1. Select all the objects that you intend to color during this **Live Paint Bucket** session.
2. Choose **Live Paint Bucket** (*K*), which is located under the **Shape Builder** tool.

3. You should now see your cursor change to the **Live Paint Bucket** icon along with a message stating, **Click to make a Live Paint group**. A swatch selection bar will also be displayed directly above your cursor.

4. Using the *left* and *right* arrows on the keyboard, you can move through the current swatch library. It will then continue into the next available swatch library. Using the *up* and *down* arrows, you can travel through any swatch libraries currently in your **Swatches** panel.

5. After you have selected the desired color, click on the area you would like to fill (see *Figure 6.4*):

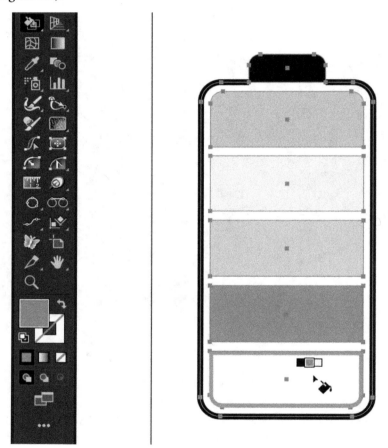

Figure 6.4 – Adjusted colors with Live Paint Bucket

6. Continue to use your arrow keys to select the next color you would like to fill with, and then click into the next area you would like to fill. You will notice that the **Live Paint Bucket** tool will offer an indicator color outline (see *Figure 6.5*):

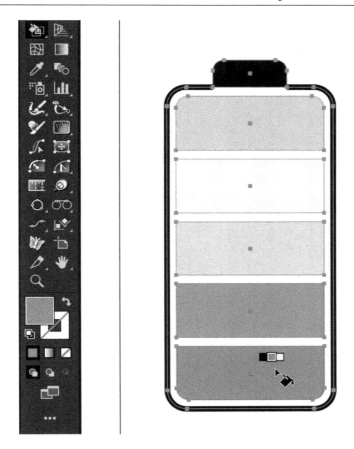

Figure 6.5 – Filled with targeted color

7. When you have colored all the available shapes and no longer need the **Live Paint** group, you can select it (if not still selected) using the **Selection** tool (*V*), and then go to **Object** > **Live Paint** > **Expand**. This will once again make it a traditional **group**. Right-clicking on this group will allow you to **ungroup** and continue to adjust their appearance further if needed.

> **Important Note**
> Once a **Live Paint** group has been created, it no longer needs to be selected before using the **Live Paint Bucket** tool. The **Live Paint Bucket** tool will not select the shape you click on and color it automatically.

But, now you might be asking yourself, "What if I wanted to have a solid fill color *and* a pattern fill?" You can do that (and more)! Let's look at how this can be done next.

Multiple fills and strokes through the Appearance panel

Using the **Appearance** panel is crucial to becoming more efficient with your process of creating within Illustrator. It allows you to have a visible record of the attributes assigned to any selected object, as well as quick access for adding to or editing attributes.

By utilizing opacity and blending methods, you can apply multiple fills to any object, and they will be able to have a blended appearance. You can also add multiple strokes that can also benefit from utilizing opacity and blending methods and can also be made visible due to the alignment method chosen or the width of the strokes. By staggering their alignment and/or changing their width, they will be visible beside each other, rather than directly on top of one another (see *Figure 6.6*):

Figure 6.6 – Multiple fills and strokes

To accomplish an appearance similar to what you see in *Figure 6.6*, you will need to open the **Appearance** panel and select the object you would like to create multiple fills and/or strokes for. The panel will show the current attributes associated with the object. Select either **Add New Fill** or **Add New Stroke** from the Options menu, and then customize the new item with your desired attribute. Remember that you must also adjust the opacity and/or blend mode to have the visual benefits of multiple fills.

In *Figure 6.7*, I added a gradient, but then adjusted the blend mode so that the original solid fill would be visible through the lighter values of the gradient:

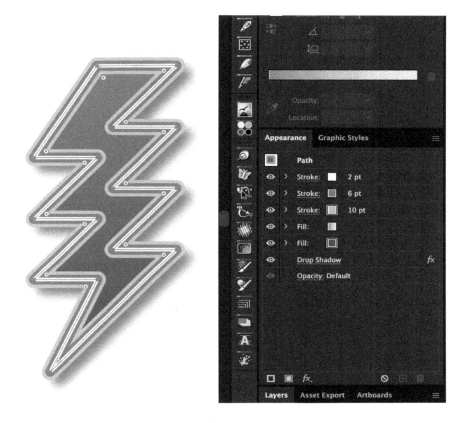

Figure 6.7 – The Appearance panel

If you apply multiple strokes to an object, you can vary their widths to create a stack (as I did in *Figure 6.7*), but you can also utilize the options of **Align Stroke to Center**, **Align Stroke to Inside**, and **Align Stroke to Outside**. These options can be found in the **Stroke** panel.

If you are looking for a way to make your strokes appear more organic or similar to traditional art media, the techniques discussed in the next section will help you to explore a variety of options to add to any stroke.

Variable width profiles and vector pack brushes for strokes

There are times when you want to have a varied line width to the stroke of an object, or even apply a stroke attribute that can mimic a traditional texture, such as a dry brush or a spattering ink pen. Illustrator allows you to customize the stroke to obtain a large variety of appearances. In this section, we will look at how to accomplish both techniques.

Variable width profiles

The **Width** tool (*Shift + W*) allows you to vary the thickness of the stroke along any path. Click and pull on any area of a path and you will be able to enlarge or reduce the width for that portion of the path. Each anchor along the path that is adjusted can be further adjusted at any time through the use of the anchors that were created to make this width adjustment. In *Figure 6.8*, you can see that the selected anchor is now assigned two additional anchors for adjusting the width:

Figure 6.8 – The Width tool used to create several adjustment anchors

To adjust the width further, just click and pull either the center anchor or one of these two additional anchors, and pull outward to thicken or inward to thin the stroke. If you click and pull the center anchor along the path, you will be able to move this width up or down the path location.

This tool allows you to develop a random stroke that shrinks and swells at any point you choose. It is excellent for giving you greater control of a path that can benefit from a varied width. I find that it allows you to customize your strokes for a more hand-drawn look, which takes away a bit of the predictability of an even stroke in your illustration.

In *Figure 6.9*, I adjusted the stroke of the two leaves to allow for the appearance of a stronger downward brushstroke at the bottom of each leaf:

Figure 6.9 – Variable width applied to strokes on leaves

Even if it is only a small or subtle change (as is the case in *Figure 6.9*), it will still be a benefit to your illustrations if you are looking for ways to make your illustrations stand out from others.

Vector pack brushes

Two vector packs come with Illustrator: the **Grunge** and **Hand Drawn** vector packs. Of course, you can create your own vector brush from any vector object or add any already created vector brush after acquiring them elsewhere.

To add a vector brush from either the **Grunge** or **Hand Drawn** libraries, follow these steps:

1. Open the **Brushes** panel.
2. Access the **Brushes Library** menu in the upper right or lower left of this panel.
3. In the new window, locate the **Vector Packs** group and then choose from one of the two vector **Brush Libraries**.
4. You will now be able to apply the added brush to any path. Select the intended path and then click the brush, which now resides in your **Brushes** panel.

After applying a brush, you may notice that it isn't quite what you were looking for. If you need to adjust the appearance of any brush, you can select it and then open the **Stroke** Options panel (see *Figure 6.10*):

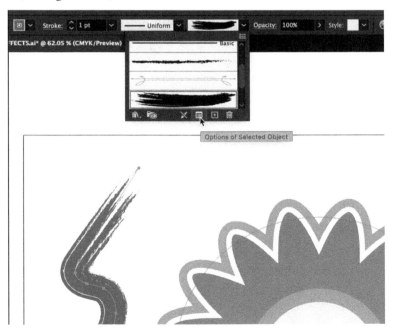

Figure 6.10 – Initiating Stroke options

From within this panel, you can obtain access to the **Options of Selected Object** button (see *Figure 6.10*), which will open the **Stroke Options** panel (see *Figure 6.11*).

The most common action I use this panel for is to resize the brush stroke. Of course, you have several additional editing choices here, including the ability to flip the brush stroke:

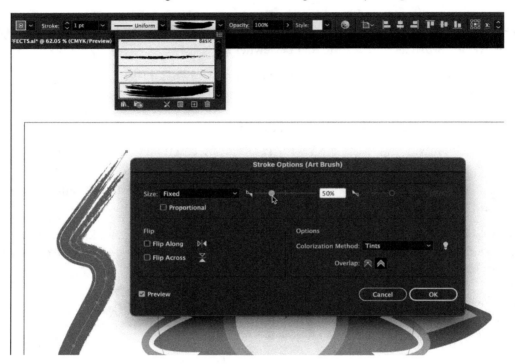

Figure 6.11 – Adjusting the scale of the brush stroke

You can also control the technique that Illustrator uses for the overlap of the brush stroke at corners (**Overlap**). You can set it to **Adjust corners and folds to prevent overlaps** (the highlighted button lower right of *Figure 6.11*), so the brush maintains the look of the brush stroke without having it overlap with itself. If you choose **Do not adjust corners and folds**, the stroke will often have an odd overlapping behavior at the corners, which will result in the brush stroke design appearing to go off the intended path and becoming transparent wherever it overlaps itself. This usually happens more as you widen the stroke.

Next, we are going to look at some methods that could be excellent time savers, especially if you need to make quick changes to the same symbols often.

Dynamic tools for greater options

As this chapter is about advanced attribute design, I would be amiss if I did not review some of the more advanced options for creating within Illustrator. These methods have been available since 2015, and have allowed designers greater flexibility when it comes to creating, adjusting, and adapting artwork. We will be looking at the following three ways in which you can have greater flexibility in your decisions:

- The **Shaper** tool (*Shift + N*)
- **Live shape** options
- **Dynamic Symbols**

Now we will discuss all three techniques, which offer dynamic options to assist you in creating or editing an object with the assistance of the software.

Shaper Tool

The **Shaper** tool (*Shift + N*) allows another technique for adding basic shapes to the artwork. Using a mouse, drawing tablet, or touchscreen, you can create a gesture of the intended shape, and Illustrator will attempt to interpret what you intended and then replace your gestures with the basic shape.

Live shape options

The ability to adjust a shape after it has been created using an assortment of on-shape **control widgets** is a method that Adobe has dubbed **live shape editing**. It allows you to dynamically adjust the shapes without the need for changing the tool or using any effects or filters. You can also make use of the controls available in the **Transform** panel to make live shape adjustments, as shown here:

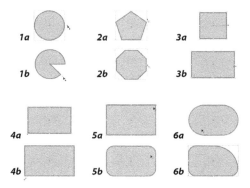

Figure 6.12 – Use of control widgets available on live shapes

Let's review *Figure 6.12* for methods of editing a live shape:

- Example 1 illustrates the pie widget icon (**1a**) and its use to create a pie shape (**1b**).

- Example 2 illustrates the option to add or remove sides to a polygon (**2a**). Click and drag the icon to either add or remove sides (**2b**).

- Example 3 illustrates the use of one of the eight bounding box handles (**3a**) to scale a shape. Using a side handle will scale the shape in one direction (**3b**).

- Example 4 illustrates the use of a corner bounding box handle (**4a**). Using a corner handle will scale the shape proportionately (**4b**).

- Example 5 illustrates the corner widget icon (**5a**) and its use in the corner radius (**5b**).

- Example 6 illustrates that when the radius preview is red, it means that it has reached its maximum possible limit (**6a**); while using the **Direct Selection** tool, you can select individual corners and then adjust the radius (**6b**).

In addition to all the aforementioned uses of the control widgets, live shapes can also be rotated by moving your cursor slightly beyond the bounding box corner. Your cursor will change to a bent arrow icon, and then you can click and drag to rotate the shape.

Dynamic Symbols

The **Symbols** library offers a great benefit for your creative workflow. It is a great place to save elements that you intend to use often or could benefit from slight adjustments made dynamically. First, let's review what steps to take when editing a symbol so that all incidents of the symbol change:

1. After opening the **Symbols** library, select the symbol you would like to use in your work, and then choose **Duplicate Symbol** from the Options menu in the top right of this panel. This is not mandatory, but it will allow you to keep the original symbol as it is and then make changes to your newly created copy.

2. Drag an instance of the symbol onto your canvas. You can add more with this same technique, or hold the *Alt/Option* key while dragging the first instance with the **Selection** tool to make additional instances of the symbol (see *Figure 6.13*):

Figure 6.13 – Symbols from the Retro library

3. Using **Selection** tool (*V*), select any instance of the symbol and then adjust it by double-clicking it, and clicking the **Edit Symbol** button located in the Options bar or through the **Recolor Artwork** panel (also in the Options bar).

4. Select the element you would like to change and then adjust it through the **Properties** or **Appearance** panels.

5. Double-click outside the symbol instance to exit the **Edit Symbol** isolation mode.

Adjusting any element of one of the symbols will automatically adjust all incidences of this symbol. I have found the **Recolor Artwork** icon located on the Options bar (color wheel icon), the **Recolor** button in the **Quick Actions** section of the **Properties** panel, or you can go to **Edit** > **Edit Colors** > **Recolor Artwork...** from the top menu, which will give you access to adjust colors through many methods (*Figure 6.14*):

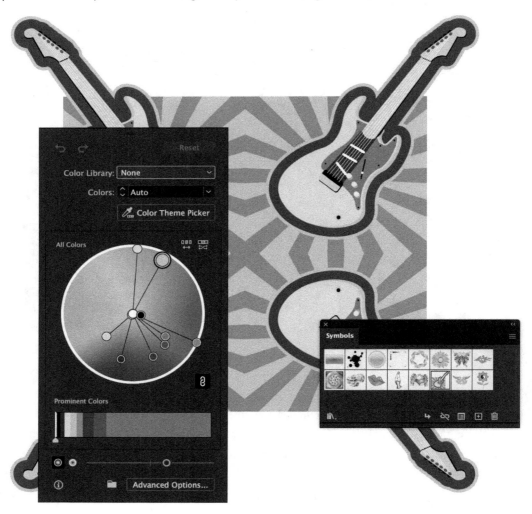

Figure 6.14 – Adjusting a symbol with the Recolor Artwork tool

Although the new panel for the **Recolor Artwork** tool has much of what you would need, you can also choose the **Advanced Options** button, located in the lower right of the panel, to access all controls that are available with the **Recolor Artwork** tool. This is the legacy panel for this tool, so if you are already familiar with **Recolor Artwork**, you will likely appreciate the option to still access this. In *Figure 6.15*, you can see the added choices in the **Advanced Options** panel:

Figure 6.15 – Adjusting a symbol with the advanced recolor options

After editing a symbol in *Figure 6.16*, you can see that all instances, as well as the symbol in the **Symbols** library, have been updated:

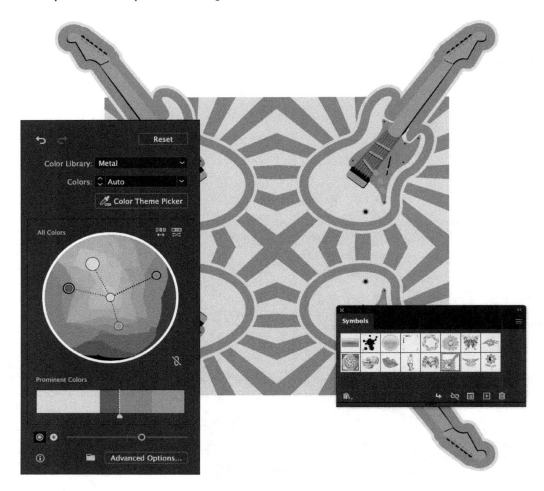

Figure 6.16 – All symbol instances automatically update

By changing the symbol to a dynamic symbol, you can change the symbol and all instances for it, or for just the one instance of the symbol you are currently adjusting. To change an individual instance (or an individual element of it) follow these steps:

1. After opening the **Symbols** library, select the symbol you would like to use in your work, and then choose **Duplicate Symbol** from the Options menu in the top right of this panel. This is not mandatory, but it will allow you to keep the original symbol as a static symbol.

2. Select the duplicated symbol, then choose **Symbol Options…** from the Options menu, and check the **Dynamic Symbol** button on the **Symbol Options** panel. This will result in having a + sign added over the symbol in the library view (see *Figure 6.17*):

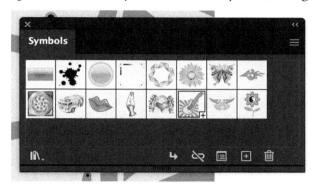

Figure 6.17: Dynamic Symbol indication

3. Using the **Direct Selection** tool (*A*), select any element of the symbol (or hold *Shift* while selecting multiple elements) and then adjust it through the **Properties** or **Appearance** panels (*Figure 6.18*):

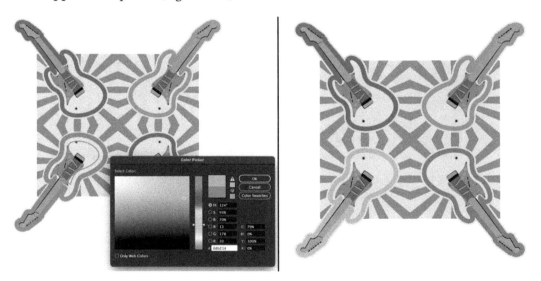

Figure 6.18 – Dynamically editing individual elements of symbol incidences

Adding the ability to dynamically edit a symbol is perfect for those situations where your symbol can be used as a template. For example, a fashion designer might place a symbol for a shirt that will need to be presented in several colors.

Although we have just talked quite a bit about color and different ways to make adjustments to it, I love that there are still even more options for how a user can work with color.

Adobe Color offers a myriad of options for expanding both your color knowledge and color options. You can begin by going to `color.adobe.com` and logging in using your Adobe account. From there, you will be introduced to the site from the **Create** tab. Here, you will find several options for working with color:

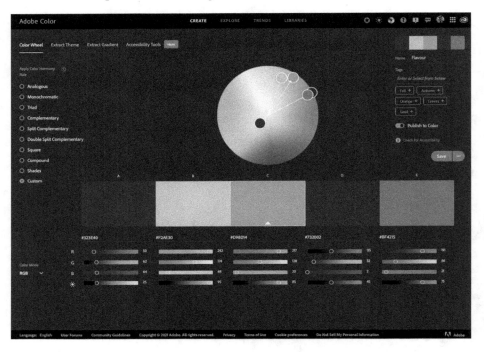

Figure 6.19 – User interface of Adobe Color

Adobe Color allows for use of color harmony rules (color theory), extracted color palettes (from any image), and curated palettes (industry trends). You can save any palette to your **Creative Cloud library** and then access it within Illustrator (and other Adobe software titles). Color harmony rules are located at the top left of the page (*Figure 6.20*):

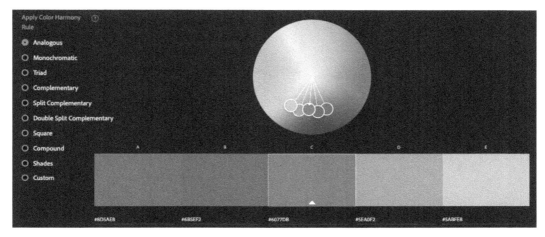

Figure 6.20 – Analogous color harmony rule applied in Adobe Color

A palette allows for five colors and will present the **hex color code** directly underneath each swatch. Approved palettes can be saved for later use in the **Creative Cloud library** of your choice.

Summary

Adobe Illustrator offers so many ways to create and adjust the appearance of vector paths and the shapes they create, that it may feel like a daunting task to decide which tools and techniques to use for each situation. But, from observing these options to the appearance attributes of a path, you should have some new ideas for customizing the look of any shape you create. You might become more astute at the use of colors and color harmonies. You might take this previous experience and use it to make even wiser choices on changing the appearance of an illustration. It should become easier for you, over time, to make choices regarding the options for texture, pattern, and color attributes.

In the next chapter, we are going to consider the options and benefits of text in your designs and apply them to a sample illustration.

7

Powerful Typography Options in Adobe Illustrator

Now that we have explored a variety of methods to create and adjust the appearance of objects, we are next going to look at several methods of creating and adjusting your **typography** in a design. We will be looking into the components of a font, as well as how different appearances formulate a mood or meaning to assist you in your intended communication. To accomplish this, the chapter will be divided into the following main topics:

- Exploring font structure and personality
- Using ligatures and alternatives
- Designing with text on a path
- Drawing using Snap to Glyph

Technical requirements

To complete this chapter, you will need the following:

- Adobe Illustrator 2022 (version 26.0 or above).

- High-quality internet access may be required for some situations.

- Access to Adobe Fonts (`https://fonts.adobe.com`) through a current Adobe account will allow you greater access to quality fonts.

Exploring font structure and personality

Selecting the right font is a crucial element in designing quality work that communicates the intended mood or meaning to the viewer. Although nothing is a substitute for experience, adding additional knowledge related to the art of typography will empower you to make wise design choices as you continue to hone your skills with fonts. Let's look at some terms used for describing the anatomy of a font, and then move on to suggest some methods for utilizing type and its options. To understand the components of a font, I recommend an excellent online resource called Fontshop which has an excellent glossary page (you can view this here: `https://www.fontshop.com/glossary/`).

The key point to remember is that fonts are designed with different characteristics to elicit a response from the viewer. Each design when applied to the text gives it a certain personality. In this section, we will look at how you can make a choice of font and add any desired adaptations needed to improve the font's intended communication.

To customize the text, you will need to become familiar with two panels that will be crucial for the task. From the top menu, choose **Window** > **Type** > **Character** or **Paragraph**. From within the **Character** and **Paragraph** panels, you will find options to adjust the typography within Illustrator:

Figure 7.1 – The Character panel

Let's first look at the **Character** panel and its uses, and then we will examine the options available to you in the **Paragraph** panel.

The following list is intended to break down the meaning and use of each button in the **Character** panel:

- **Set the font family**: In this top box, you can select from all installed fonts.

- **Set the font style**: In this box, you can choose between available style options for the chosen font (**Regular**, **Bold**, **Italic**, and more).

- **Set the font size**: Choose from a list of available font sizes from **6** to **72 pt** or type in any desired size. In addition, you can use the up arrow to increase or the down arrow to decrease the size of the font.

- **Set the leading**: The leading (pronounced "ledding") is the space between two or more lines of text.

- **Set the kerning between two characters**: To adjust the space between two characters, or **kern** them, you need to first place the insertion cursor between the two characters and then use the *right* and *left* directional arrows on your keyboard while holding down *Alt/Option* to adjust the space between them. The *right* arrow will widen, while the *left* arrow will narrow the space.

- **Set the Tracking for the selected characters**: Adjusting this will allow you to loosen or tighten the space between all the characters within a selection.

- **Vertical Scale**: Illustrator can adjust elements of the font that are not available directly within the font family. **Vertical Scale** allows you to stretch the characters vertically to create custom characters that are taller than the original font.

- **Horizontal Scale**: This allows you to stretch the characters horizontally to create custom characters that are wider than the original font.

- **Set the baseline shift**: This allows you to have characters that have an adjusted baseline, which will adjust them to become either a subscript or superscript element.

- **Character Rotation**: Select one or more characters and then select a rotational angle. All the angles are in relation to the original baseline.

- **Faux Adjustments**: Although these options are not in the original font families, they can be applied using Illustrator's ability to create adjustments to the chosen characters, and can also be used in combination:

a. **All Caps**: This will bring all text to the cap height of the font. Here is an example:

ADOBE ILLUSTRATOR FOR CREATIVE PROFESSIONALS

Figure 7.2 – All Caps creates consistent font height

b. **Small Caps**: Illustrator can create small caps using scaled-down versions of the regular capital letters. Here is an example:

ADOBE ILLUSTRATOR FOR CREATIVE PROFESSIONALS

Figure 7.3 – Small Caps replaces lowercase type

c. **Superscript**: This is also known as **superior** text. It is reduced-size text that is raised in relation to the font's baseline. Here is an example:

Adobe Illustrator for Creative Professionals, 1st Edition

Figure 7.4 – Superscript used to create placement number

d. **Subscript**: This is also known as **inferior** text. It is reduced-size text that is lowered in relation to the font's baseline. It is most often used for mathematical and chemical formulas. Here is an example:

The formula for water is H_2O

Figure 7.5 – Subscript used to create chemical formula

e. **Underline**: A line can be made below the baseline of any text. The default weight of an underline will depend on the size of the text. Here is an example:

<u>Adobe Illustrator for Creative Professionals</u>

Figure 7.6 – Underline used to emphasize type

f. **Strikethrough**: A line can also be made that goes through the center of the *x*-height of the text. The default weight of a strikethrough will depend on the size of the text. Here is an example:

~~Adobe Illustrator for Creative Professionals~~

Figure 7.7 – Strikethrough used to communicate edit

- **Language**: In this box, you can choose a language or a regionally specific version of a language. This can be utilized when engaging **Spell Check** for any document. You will find this by going to the top menu and choosing **Edit** > **Spelling** > **Check Spelling**.

- **Set the anti-aliasing method**: Illustrator offers you several choices for smoothing the edges of the characters when preparing art for bitmap output.

- **Snap to Glyph**: Adobe has recently added this updated feature, which allows you to align objects in relation to elements of type (a **glyph** is defined as a written or printed mark that is meant to convey information to the reader). Illustrator allows for alignment to the following elements related to type:

a. **Baseline**

b. **X-height**

c. **Glyph Bounds**

d. **Proximity Guides**

e. **Angular Guides**

f. **Anchor Point**

In *Figure 7.8*, you can see an example of adjusting the text using the **Character** panel:

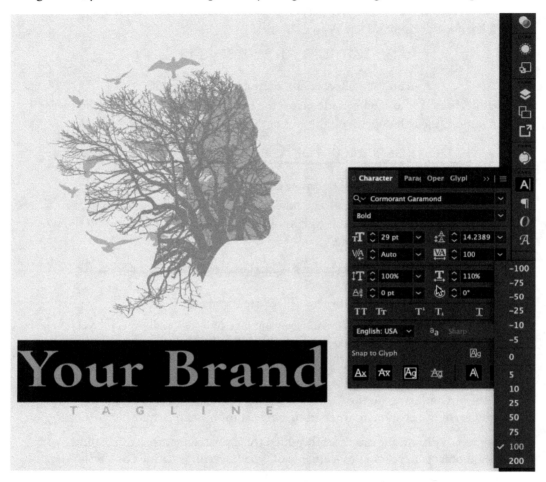

Figure 7.8 – Adjusting tracking

Note that the space between the letters of the highlighted text is being expanded by choosing **100** in the **Tracking** portion of the panel.

In addition to all the adjustments the **Character** panel allows, you can also adjust blocks of text using the **Paragraph** panel:

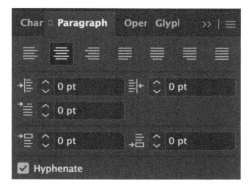

Figure 7.9 – The Paragraph panel

The **Paragraph** panel allows you to achieve greater control of spacing and structure within paragraphs of text. You will find all the following options within the **Paragraph** panel:

- **Align Left**: All text will be aligned to the left margin.

- **Align Center**: All text will be aligned between the margins.

- **Align Right**: All text will be aligned to the right margin.

- **Justify with last line aligned left**: All text will be spaced to fill each line while being aligned to both the left and right margins (justified), except the final line, which will be aligned left.

- **Justify with last line aligned center**; All text will be spaced to fill each line while being aligned to both the left and right margins (justified), except the final line, which will be aligned center.

- **Justify with last line aligned right**: All text will be spaced to fill each line while being aligned to both the left and right margins (justified), except the final line, which will be aligned right.

- **Justify all lines**: All text will be spaced to fill each line while being aligned to both the left and right margins (justified) including the final line.

- **Left indent**: This setting allows you to indent a paragraph a certain amount away from the left margin.

- **Right indent**: This setting allows you to indent a paragraph a certain amount away from the right margin.

- **First-line left indent**: This setting allows you to indent the first line of a paragraph a certain amount away from the left margin.

- **Space before paragraph**: With this setting, you can create a certain amount of space before a paragraph.

- **Space after paragraph**: With this setting, you can create a certain amount of space after a paragraph.

- **Automatic Hyphenation**: This allows Illustrator to hyphenate words appropriately instead of only returning words to the next line when there is not enough space on the current line.

After mastering the **Character** and **Paragraph** panels, you will be well on your way to gaining complete control of your text. At times though, you will want an option with extra flair, or an option based on a historical context. We will discuss these options in the next section.

Using ligatures and alternatives

Most print fonts will be listed as one of two types: **TrueType** or **OpenType**. At this stage in history, pure TrueType fonts are considered outdated and are no longer being made, so adding these older font types is generally not recommended. For more information on the history of font types, check out this link: `https://helpx.adobe.com/in/illustrator/using/fonts-old.html`.

Ligatures and **alternatives** are available within the **OpenType** font options. The term ligature comes from the Latin term *ligare*, which means "to tie." Some fonts offer ligatures for joining two or three common combinations of letters. They can be used for maintaining proper kerning or for decorative purposes. Alternatives allow you to customize the look of your text by selecting from a selection of alternative character choices.

Figure 7.10 shows the many options available for replacing characters or common character groups:

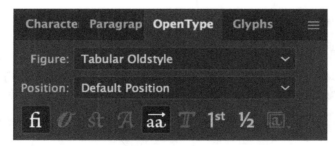

Figure 7.10: The OpenType panel

The **OpenType** panel offers several options for those OpenType fonts that have been designed to take advantage of ligatures and alternatives:

- **Figure**: Used for alignment of numerals.

- **Position**: Used for positioning fractional, subscript, and superscript numbers.

- **Standard Ligatures**: Combines awkward letter groupings (such as **fi**, **fl**, **ff**, **ffi**, and **ffl**) into a more elegant appearance.

- **Contextual Alternates**: Ligatures applied to individual characters based on the letters around them.

- **Discretionary Ligatures**: Ligatures that may be used at the user's discretion.

- **Swash**: A flourish or exaggerated serif.

- **Stylistic Alternates**: Available prior to the inclusion of **Stylistic Sets**.

- **Titling Alternates**: Specially designed for large, all-caps titles.

- **Ordinals**: Letters used to represent placement (1st, 2nd, and so on).

- **Fractions**: Used to create the diagonal appearance of fractions.

- **Stylistic Sets**: If the font has multiple sets, you can select from them to find offered alternatives.

Figure 7.11 illustrates the use of a standard ligature in place of the "ff" in the word "coffee":

Figure 7.11 – Example of a standard ligature

To select alternatives for the font, highlight any character and then hover over it. If available, any alternatives will display directly below it. If several alternatives are available, there will also be an arrow to the right. Click it to open the **Glyphs** panel where all available alternatives will be displayed (see *Figure 7.12*):

Figure 7.12 – Alternatives presented in the Glyphs panel

Now that we have examined the opportunities that **Type** and its options can offer you, let's look at a great option for type placement. There's more than the traditional textbox to choose from, and these additional methods allow for a lot of design opportunities.

Designing with text on a path

Text can be directly written upon any path or can be pasted from any traditional text frame. **Type on a Path** Tool is in the **Type Tool** family of tools in the toolbar. Click and hold the icon of any tool within this family to access the additional tools, including **Type on a Path Tool**. Choosing this tool will now adjust the cursor to a variation of the insertion cursor you usually see when choosing the regular **Type Tool** (see *Figure 7.13*):

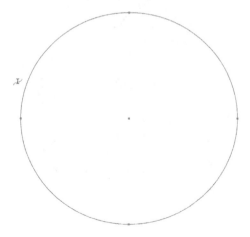

Figure 7.13 – The Type on a Path cursor

In addition to **Type on a Path Tool,** there is also **Vertical Type on a Path Tool.** It works very similarly to **Type on a Path Tool.**

To apply text to a path, you can just move the cursor over the selected path and then click on it. It will then be filled with highlighted placeholder text. At this stage, you can change the highlighted placeholder text by typing the intended text. The text can then be modified and adjusted along the path (*Figure 7.14*):

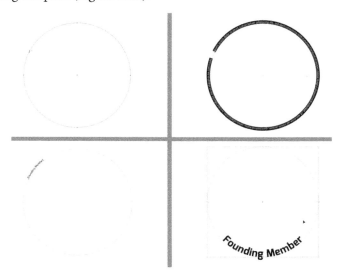

Figure 7.14 – Customizing the text along the path

You can center the text or move it anywhere along the path by selecting and moving the center bracket located in the middle of the text. Pulling down while moving this bracket will also allow you to flip the text to the other side of the path. Hold down the *Ctrl/ Command* key to keep the text from flipping to the other side. There will also be a left and right bracket, which allows you to control the margins of the textbox. These can be used to ensure that your text is centered.

In *Figure 7.15*, I have placed the left bracket on the left anchor of the circular path that the text is on, and then placed the right bracket on the right anchor that the text is on:

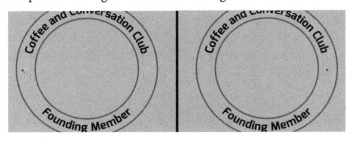

Figure 7.15 – Centering text on a path

This allows you to then choose **Align Center** from the **Paragraph** panel to center the text along the arc of the circular path the text resides on.

Adjust your design with additional icons, fill and path adjustments, and patterns. Try out several fonts to find that perfect mood for your design. Be sure to consider design principles as you develop the work. Is it unified? Does it have an area of emphasis? Does it create a formal or informal balance? If something feels out of place and doesn't fit in, then you are getting there, but are just not yet done. Keep working to get it to come together as one cohesive design, as the progression here shows:

Figure 7.16 – Designing with text on a path

The jump from *Figure 7.16* to *Figure 7.17* illustrates the need to continue to work to find a resolution that has better harmony. The font, stroke widths and icon stroke width were all adjusted to develop a design that has a great sense of unity, as follows:

Figure 7.17 – Additional edits

An additional option in the same family of text tools is **Area Type Tool**. It will allow you to fill text into any shape. I have included an example of it in *Figure 7.18*:

Figure 7.18 – Area Type Tool used for background

Note that to use **Area Type Tool**, like **Type on a Path Tool**, you must click on the path of the intended area. The **Paragraph** panel options for alignment can also be used on the text applied with **Area Type Tool**.

You will need to experiment with this a bit, as simplified shapes generally work better than more complex ones with **Area Type Tool**, but can be used when fonts are small.

Next, we are going to take a closer look at the Snap to Glyph option that was discussed earlier in this chapter.

Drawing using Snap to Glyph

With the addition of this more recent option, you now have several additional techniques for alignment related to type. If you already make use of **Smart Guides**, then this will feel very familiar to you. If you are not yet using **Smart Guides**, I would first recommend you explore their use. You will find them by going to **View** > **Smart Guides**. This will turn on interactive guides that will assist you with the alignment of objects. Similarly, **Snap to Glyph** will allow you to align (or **snap**) elements to different parts of your type (**glyphs**).

The **Snap to Glyph** options are located at the bottom of the **Character** panel:

Figure 7.19 – Snap to Glyph options

Figure 7.19 shows the options in icon form. The following list helps to describe these options from left to right:

- **Snap to a specific glyph**: Right-click on a glyph and select **Snap to Glyph** to view glyph-based guides. After creating or moving an element in relation to the specific glyph, select the **Release Glyph** button located in the **Type** option bar.

- **Baseline**: Snap to the base of glyphs while drawing, moving, or scaling.

- **x-height**: Snap to the height of lowercase (for example, x,z,v, or w) glyphs while drawing, moving, or scaling.

- **Glyph Bounds**: Snap to top, bottom, left, or right bounds of glyphs while drawing, moving, or scaling.

- **Proximity Guides**: Snap to guides generated near the baseline, x-height, and glyph bounds while drawing, moving, or scaling.

- **Angular Guides**: Snap to angles of a glyph while drawing, moving, or scaling.

> **Note**
>
> **Angular Guides** appear when the textbox is rotated or when a specific glyph with angular segments is selected from the right-click menu.

- **Anchor Point**: Snap to anchors of a glyph while drawing.

Figure 7.20 shares the creation of an Illuminated letter created by using the **Snap to Glyph** tools. The background and framing were created using the **Glyph Bounds** option, while the inner pink shape was created using the **Anchor Point** option:

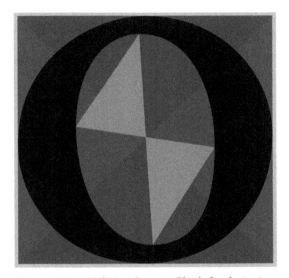

Figure 7.20 – Utilizing Snap to Glyph for designing

Utilizing the **Snap to Glyph** tools can offer many design options while creating pieces where text is a strong element in the piece. It offers greater precision and speed when compared to placing guides and then snapping content to them. It may take a bit of practice, but I'm confident that you will love **Snap to Glyph** more as you use it.

Summary

Of course, we have only scratched the surface when it comes to the use of typography and all the advanced methods that type offers, but this chapter was intended to increase your knowledge of typography and motivate you to continue to learn to control your type.

In this chapter, we have reviewed the options available for **Type** from within the **Character** and **Paragraph** panels, including the **Snap to Glyph** options. We looked at ligatures and alternatives that can be used to give more options for customizing your text. We also looked at the ability to apply text to a path using the **Type on a Path Tool**.

Remember that your text will assist in the communication and feeling of your work. All that we learned in this chapter is near useless if you aren't applying it with the intent to bring out the appropriate mood or meaning from within your work. Sometimes, it is as subtle as the style or size of the text, while at other times it is a bigger decision, such as where text is placed or the spacing of characters. All these tools allow you to make these adjustments, but ultimately, it will be your design sense and intuition that allow you to adjust the text to elicit the intended personality of the piece.

In the next chapter, we will be looking at the many ways for preparing your work for presentation and viewing.

Part 3 – Real-World Applications

When you have gained new knowledge and skills, this part will help you to prepare more complete presentations and compositions suitable for clients.

This part comprises the following chapters:

8
Preparing Artwork for Presentation

In all the remaining chapters, we will be working together to create artwork that could be considered suitable for presentation to a variety of audiences. This chapter is intended to guide you through several design scenarios. Each will require a different output method based on its intention. We will look at standard methods for preparing illustrations from original artwork. Then, we will look at examples of a couple of popular industry presentations. We will also discuss all the various file formats that can be used and how to export them for your intended media.

To accomplish this, the chapter will be divided into the following main topics:

- Starting with the end in mind
- Creating industry-standard presentations
- Output methods for more media options

Technical requirements

To complete this chapter, you will need the following:

- Adobe Illustrator 2022 (version 26.0 or above).

- High-quality internet access may be required for some situations.

- Adobe Stock.

Starting with the end in mind

In my humble opinion, you are doing yourself a disservice if you don't use the traditional method of paper and pencil to begin your concept phase.

Drawing out your concept first can often allow you to accelerate the ideation process. In addition, it can be used as a template to create from throughout the entire design process. Use the following steps to check your understanding of this process. We will begin by planning out an illustration, and then develop it on top of the sketch's layer after making it a template layer:

1. **Start with the plan**: Consider the intended output for your work. Although it isn't essential, I would recommend you start with a medium that is either the exact size of the intended output or a ratio of it. This will allow you to easily scale the artwork to the file's output size without distortion. For example, if you intend to print a large file at 16x20 inches, then draw on a 4x5 inch or 8x10 inch piece of paper. Likewise, if you are using international sizing, and you intend to print a large file on A3 paper, then draw on an A5 or A4 piece of paper.

2. **Prepare your work for Illustrator**: Either scan or photograph the work and then move the file to a place on your workstation from where you can retrieve it. I recommend a project folder that is also being backed up to a cloud drive.

3. **Open an Illustrator file**: Select the appropriate media and document size.

4. **Place your artwork**: Go to **File** > **Place…**, then locate your artwork and hit **Place** (or simply double-click on the artwork). For this sample project, you can start with any original drawing or photograph (see *Figure 8.1*):

Figure 8.1 – Original artwork placed into an Illustrator file

5. **Size your work to the file**: If you created the artwork on the correct sized medium (which means 8x10 inch paper for an 8x10 inch print), place the cursor in the top-left corner of the artboard boundary so that the top and left guides of the icon line up with the guides of the artboard. Right-click the mouse and you should find the artwork is placed within the document's artboard.

    If you created the artwork at a certain ratio (which means 4x5 inch paper for an 8x10 inch print or A5 for an A4 print), perform the same initial step of placing the cursor in the top-left corner of the artboard boundary so that the top and left guides of the icon line up with the guides of the artboard, but this time, click and pull the icon until your artwork is scaled to the document's artboard boundary.

6. **Dimming your artwork**: Go to the **Layers** panel and make sure that the image is your active layer. Choose **Template** from the **Options** menu (icon at the top right of the panel). This will adjust the *visibility* icon of the layer to a *template* icon, as well as lock the layer. In addition, this layer will be dimmed to 50% opacity by default (see *Figure 8.2*):

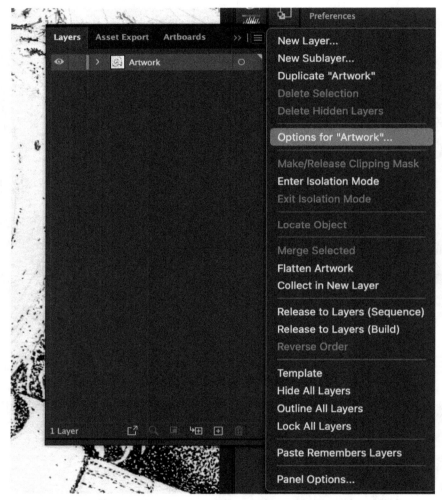

Figure 8.2 – Selecting the Options menu to change the layer to Template

In this same dialog box, you can also adjust the amount by which the placed image will be dimmed. I would recommend choosing an amount that gives the illusion of tracing paper and creates a balance between seeing your original placed artwork and the new vector paths you are creating on top of it. This dimming amount can be changed at any time (see *Figure 8.3*):

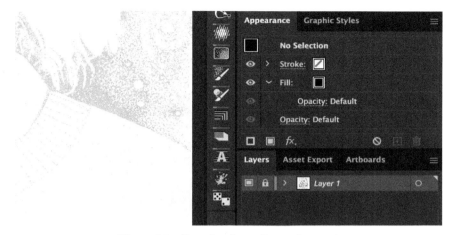

Figure 8.3 – Template layer dimmed to 35%

7. **Layer Management**: Create a new layer to develop your artwork in. Although not required, I highly recommend naming the layer at this time. Of course, it can always be done later in the **Layer Options** panel, or you may choose not to label the layer at all. Remember that naming the layers is intended to help you keep track of and make sense of your work. The **Layers** panel will never be seen by viewers of your finished work, so choose what works best for you. In *Figure 8.4*, I simply added BG to remind me to only add background material into this layer:

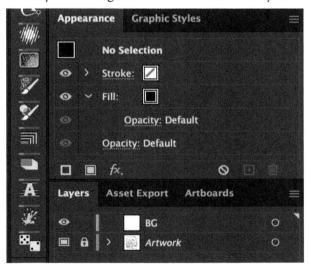

Figure 8.4 – Adding additional layers

Another management method you can take advantage of in the **Layer Options** panel is the ability to select an identifying color for the layer. It will represent the active layer with this color in the **Layers** panel and highlight all the selected vectors in that layer with this color too (see *Figure 8.5*):

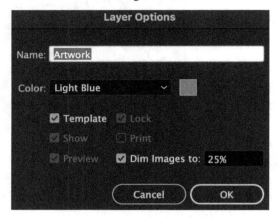

Figure 8.5 – Layer Options including a layer name input box and identifying color

8.   **Working with View modes**: To create this illustration, we are going to jump between the **Preview** and **Outline** views quite a bit. The keyboard shortcut for this is *Ctrl/ Command + Y*, and it really helps, as we will be going back and forth often to create this illustration. The **Preview** mode allows you to see the vector artwork with all attributes, while the **Outline** mode only shows the vector paths and allows you to view your originally placed artwork in a much better way. *Figure 8.6* illustrates this benefit well:

Figure 8.6 – Outline (L) and Preview (R) modes

Remember, the shortcut will show you whichever viewing mode you are not currently in, and will then bring you back to the opposite viewing mode by applying the shortcut again.

9. **Beginning your artwork**: To complete an illustration like this Einstein example, try using only **Pen Tool** and only creating straight lines with it. To keep all the vector lines straight, you just need to click for each anchor you apply. Two additional tips, I would share here are as follows:

a. Zoom in when working on details or when vectors overlap each other, as it will offer you greater control.

b. Develop layers by themes to have better control of more complex work.

In this example, you can observe that I have created a layer for the placed art, background, shirt, face, and details. Locking layers will allow you to restrict which layers you can draw in at any given time (see *Figure 8.7*):

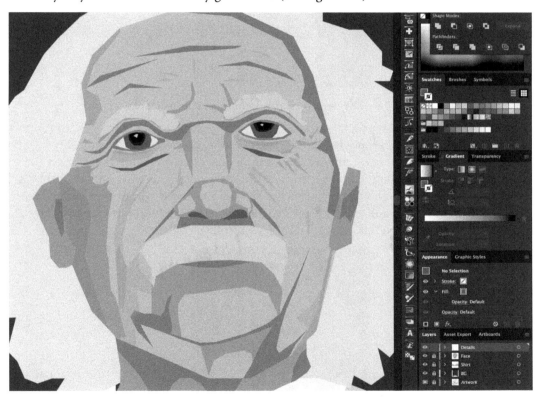

Figure 8.7 – Develop layers by themes

10. **Developing the artwork**: Continue drawing shapes and arranging them back to front to cover any overlap. *Figure 8.8* shows a further progression of the shapes being developed for this illustration of Albert Einstein:

Figure 8.8 – Developing the illustration using the Pen tool

11. **Adding final details to the artwork**: At this stage in the illustration, you will continue to develop additional shapes to replicate the shadows, highlights, and textures. Another method you may employ when getting closer to completion is to bring a copy of the original placed file to the top of the layer order. This will temporarily block the layer below. To combine the placed artwork with the vector art below it, go to the **Transparency** panel and choose a combination of **Blending Mode** and **Opacity** for the placed artwork. In this example, I chose **Multiply** and **70%**:

Figure 8.9 – Blending a copy of the original artwork

After completing a graphic depiction of one of your original images, I hope you have become more comfortable with the workflow of going back and forth between the **Preview** and **Outline** viewing modes. Over time, you will find it very helpful in completing more complex work while staying efficient with your time.

In the next section of this chapter, we will be looking at how you can prepare and process your work for a variety of intents that go beyond a single illustration.

Creating industry-standard presentations

When preparing your work for the final presentation, it is imperative that you have several items before you get started. For the two examples in the following subsections, you will be attempting to elicit a positive response from the individuals that view them. They are both intended for marketing, so you will need the images, copy, type, and colors that the client desires. Much of this can be decided through communication prior to the development of the designs.

Brand style guides

A **brand style guide** (also known as a brand identity guide, brand guidelines, or brand manual) is a document used to express the look and feel of a company's branding. The guide may be presented on screen, hosted online, or offered as a print document. Although there are many variations, the brand style guide might include the following items:

- **Mission statement**: A statement supporting the goals or direction the company intends to work towards

- **Color palette**: The limited selection of colors that will be utilized to represent the identity of the brand

- **Typography**: A selected collection of fonts that will also be utilized to strengthen the identity and maintain the consistency of the company's brand

You can see an example in *Figure 8.10*:

Figure 8.10 – Brand style guide

The design format of the brand style guide can vary quite a bit and may be presented in a variety of media types. Traditionally, brand style guides were printed, but it is now quite common to consider presenting them as a digital file that may either be shared with the client or presented onscreen during a meeting.

Comp cards

A **model comp card** is an industry-standard print product that you may be asked to create. The standard size for a comp card is 5.5 x 8.5 inches or A5 and can be in either portrait (vertical) or landscape (horizontal) orientation. The card is usually printed on thick cardstock (see an example in *Figure 8.11*):

Figure 8.11 – Model comp card

In developing the comp card, I would recommend using a bleed so that any full image or background element will extend past the edges of the card. *Figure 8.12* shows a typical document setup screen for a comp card:

Figure 8.12 – Setting up a comp card document

Generally, comp cards are still printed media, as they're intended to be handed out based on the talent that is highlighted on them. Like the brand style guide that we just discussed, comp cards could also be shared digitally if requested by a potential employer.

There are many methods of saving and sharing your work from Illustrator, so in the next section, we will look at what methods are available to you as a designer and when each could be used to help your work get the best response from those viewing it.

Output methods for more media options

Adobe Illustrator is a tool that allows you, as a designer, to develop your work for a multitude of media formats. It also allows you to easily convert your work to a different media than that which it was originally created for. It is still best practice to create your files with the output in mind. But, there are times when you will still need to change how the file will be seen.

There are quite a few file types that Illustrator will allow you to output to, so I have broken them down and discussed the benefits of each in the next sections.

Mobile, Web, Print, Film & Video, and Art & Illustration

Your creations in Illustrator can be output to several different formats depending on the needs of the hosting environment. For example, if you are developing for the web, you can create the document in **RGB Color Mode** rather than **CMYK Color Mode**. **RGB Color Mode** is an appropriate choice for anything intended for the web, as monitors and mobile device screens utilize red, green, and blue phosphors to express all colors. They do this by emitting electrons toward the phosphors, which in turn glow to produce the intended colors. **CMYK Color Mode** allows your final files to move toward the printing process without a major shift in color. The division of color information is based on the industry standard for printer inks. *C* is used to represent cyan, *M* for magenta, *Y* for yellow, and *K* for key. The key black is utilized to deepen the black (also known as composite black), which can be produced by mixing the other three ink colors (see *Figure 8.13*):

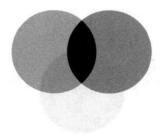

Figure 8.13 – CMYK Color Mode

Your intended media will also require you to save the file into an appropriate format. Here is a quick list (although not extensive) of many formats and their uses:

- **Joint Photographic Experts Group** (**JPEG** or **JPG**): This is one of the most common image file formats. It is ideal for web use, as it is a compressed format.

- **Tagged Image File Format** (**TIFF** or **TIF**): This is a high-quality image file format. It is intended for print use, as it is an uncompressed format.

- **Bitmap image** (**BMP**): This displays images by arranging a grid of differently colored pixels. The more pixels available within the grid, the smoother the curved edges will appear.

- **Portable Network Graphics** (**PNG**): This file format is best for reduced-color items, such as flat-colored graphics. It has the capability of selecting the bit depth to obtain the desired color definition. It also allows for an alpha channel to hold transparency data, which makes it a popular choice for web graphics.

- **Graphics Interchange Format** (**GIF**): This file format is often animated but can also be a small still image. It is ideal for web use.

- **Portable Document Format** (**PDF**): This is a format intended to maintain the original document layout and is often used to archive or share a document. An added benefit of this file type is the ability to have it signed.

- **Scalable Vector Graphics** (**SVG**): This is another file format that is intended for web use. It allows a graphic that was designed utilizing vectors to be scalable during the web design process.

- **Moving Picture Experts Group** (**MPEG-4 Part 14** or **MP4**): This is a popular video format more commonly referred to as MP4. It is a compression format, which makes it ideal for web use.

- **Cascading Style Sheets** (**CSS**): This is a style sheet language file format. It is used to communicate layout within a markup language intended for web use.

- **Photoshop** (**PSD**): As another Adobe file format, the Photoshop format works very well with Illustrator files. This is a raster-based format, so it would work well for converting your Illustrator images for screen, web, or video use.

See *Figure 8.14* for all the formats available while using the **Export** dialog:

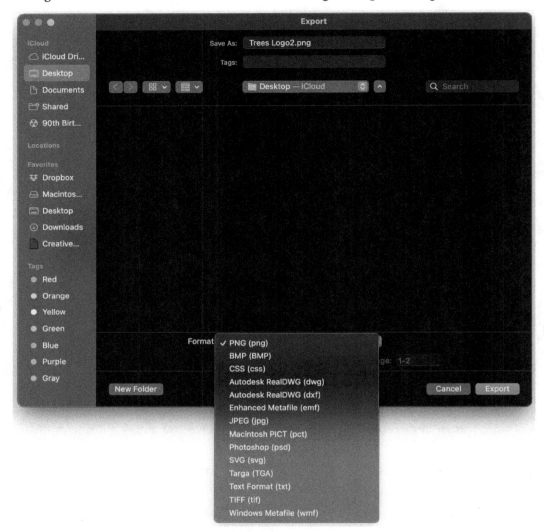

Figure 8.14 – Export dialog box and file formats

If you need to output an image to multiple formats, such as print, web, and/or video, then you should save the image multiple times. But you should also be aware that there will likely be some document changes. If you need to output an object for later use, or to use in a larger composition (a website is one good example of this), then you should also become familiar with the technique of **Asset Export** that Illustrator offers.

Asset Export

The **Asset Export** panel will allow you to move any object into this panel and save it as an independent element. Follow these steps to save an object from an Illustrator file:

1. Select every part of an object that you will be sending out via **Asset Export**.

2. Group all the shapes that were selected in step *1*. You can do that by going to the menu and choosing **Object** > **Group**. Other methods include right-clicking on the selected group and choosing **Group** from the menu that appears or using the keyboard shortcut *Ctrl/Command + G*.

3. Now, you can drag the grouped object into the **Asset Export** panel.

4. At the bottom of the panel, you will see two buttons. To the right is the **Export…** button, and to the left is the **Launch the Export for Screens dialog** button I would recommend you use:

Figure 8.15 – The Asset Export panel and the Export for Screens dialog box

This is an excellent method for collecting elements that you may later use in more advanced illustrations. If you would rather export an entire artboard or artboards, then the following sections have you covered.

Export for Screens

Now that you have been introduced to the **Export for Screens** dialog box, I can again refer to it for exporting entire artboards. At the top of **Export for Screens**, you will find both an **Assets** button and an **Artboards** button. Choose the **Artboards** button and you will see that you can now choose and export one or more of your artboards from the open document. Use the **Advanced settings for exported file types** button (gear icon) to select the file type you would like to export the artboard(s) as. You can choose between **PNG**, **JPG**, **SVG**, and **PDF** formats. This will allow you to save each artboard separately in additional formats that you will not find available to you under the **Save As** command.

One of the most important considerations you will have when working for clients or a greater intended audience is to deliver content of the highest quality. That will require you to use the most appropriate dimensions, file type, and resolution (if resolution-dependent). I have highlighted the choices of **Save As**, **Export**, **Asset Export**, and **Export for Screens** so that you may begin to explore each and discover the benefits and options within each. Over time, you will gain confidence in choosing which works best for each situation.

Summary

The ability to impress your viewer through a well-organized presentation is paramount for success. If you start with strong original content and then prepare it to a high professional standard, your intended audience is sure to be impressed. Knowing the appropriate dimensions and file format to save the image in will be needed to avoid any mishaps or delays when preparing to deliver the content to your intended audience.

Through this chapter, you learned a method for preparing illustrations from original artwork. You also looked at a couple of examples of popular industry presentations. Then, you viewed a list of various file formats that can be used and learned how to export them for your intended media.

In the next chapter, we will review the use of artboards and then attempt to extend your knowledge of their many uses. We will also look at some examples of professional workflows you might employ when working with artboards.

9
Utilizing Multiple Artboards

Imagine the power of keeping an entire client brand design within one file. Illustrator allows you to host several **artboards** within just one document. It also allows you to deliver multiple items, such as concepts, products, and pages, within a single shared file. Your client or collaborator will be happy that they only have one file to keep track of.

In this chapter, we will look at methods for setting up and controlling multiple artboards. We will also practice adding, subtracting, labeling, and arranging multiple artboards. This chapter is intended to extend the knowledge you have acquired through earlier conversations about artboards in *Chapter 2, Prepping for Illustrator*. To accomplish this, the chapter will be divided into the following main topics:

- Setting up a document with an appropriate preset or template
- Adding additional artboards
- Labeling, arrangement, and additional options

Technical requirements

To complete this chapter, you will need the following:

- Adobe Illustrator 2022 (version 26.0 or above).
- High-quality internet access may be required for some situations.

Setting up a document with an appropriate preset or template

A great method for saving valuable time is to utilize Adobe Illustrator's capability to host multiple artboards within one document, as well as its ability to save templates with one or more artboards.

If you feel that the project you are currently working on will be a common recurring task, then saving it as a template will be a wise decision. To save even more time, you might choose to search for a third-party template that you can start your work in. Why spend a large portion of your time on measuring out and preparing a template for specific items, such as box flats, when they have been created already? Be sure to check with your printer to see whether they have any templates available for your intended output media.

Next, let's review some ways that you can set up and begin a more advanced project that might require multiple artboards and/or a template.

A **New Document** screen will be presented after choosing **New** (*Ctrl/Command + N*) from the **File** menu or hitting the **New file** button on the **Start** screen when there are no files currently open. On the **New Document** screen, you will be presented with your first opportunity to create a document that hosts multiple artboards. *Figure 9.1* shows a selection of four artboards that will all be in landscape orientation:

Figure 9.1 – Creating a multiple-artboard document

There will be times when you don't want to have all the artboards uniformly sized or when you need assistance from a file that was already created. Let's first address how we can use files that host multiple artboards from outside resources. There are several preinstalled templates, and additional templates are available through **Adobe Stock**. Most of these offer multiple artboards for varied purposes. By using **File** > **New from Template…**, you will be able to open files from the `Blank Templates` folder installed with Adobe Illustrator.

In *Figure 9.2*, you can see a template that Illustrator offers you for preparing artwork to fit box flats for production printing:

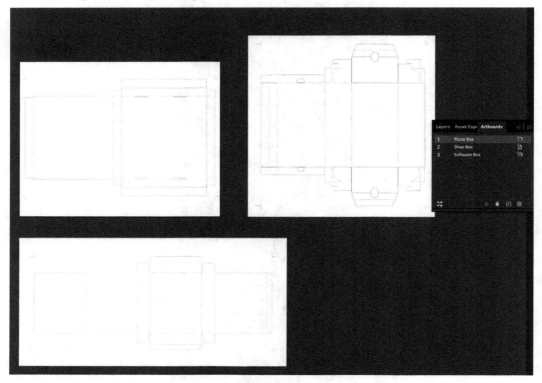

Figure 9.2 – Blank template files from Illustrator

Having multiple artboards within a single file allows you to have a variety of options with a lot less searching.

Artboards are often used to represent multiple print dimensions, such as a card and envelope combination. This technique may also be used for multiple online sizes, such as website ad banners. In this case, the designer of the template has chosen to rename all the artboards by their pixel size (see *Figure 9.3*):

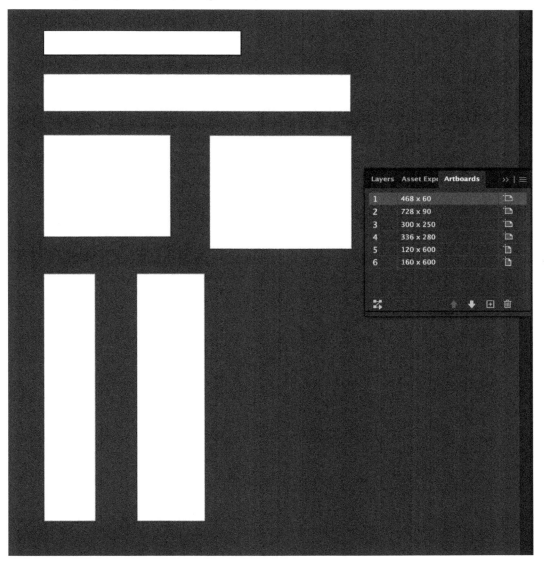

Figure 9.3 – Artboards for sizing within a template

An additional option that is available to you is to save your work as an Illustrator template file. We will start reviewing the methods for customizing artboards in the *Adding additional artboards* section, but I wanted to first point out that once you have developed a customized layout, it is easy to save it as a template that can be reused time and time again. The steps for developing a custom layout are as follows:

1. After setting up a file, go to **File** > **Save as Template…**.

2. The **Save** screen will automatically open showing the `Blank Templates` folder. Either choose to save the file there or create a new folder for your templates.

3. If you are working on an Apple computer, you might have an additional step. You might get the following dialog box:

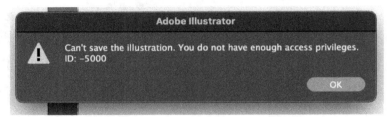

Figure 9.4 – Folder access privileges

If that happens, ensure that you are saving in the same folder: `Macintosh HD/ Applications/Adobe Illustrator CC 2022/Cool Extras /en_US /Templates/Blank Templates`.

Check your privileges on your computer for this folder (you can find this in **Get Info** or by using the *Ctrl/Command + I* shortcut).

You need the **Read & Write** permission. Change the folder permissions as follows:

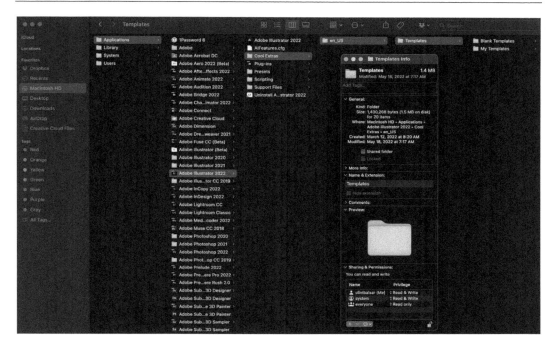

Figure 9.5 – Changing folder permissions

4. You will be able to reopen this and any other template using **File** > **New from Template…** (see *Figure 9.6*):

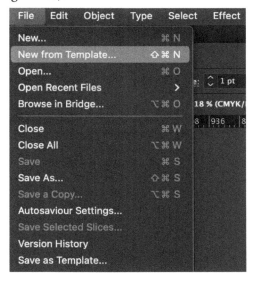

Figure 9.6 – The New from Template command

As we move into the next section, try to keep in mind that the ability to edit the current artboards and add additional artboards is available in any of the file types we have just discussed.

Adding additional artboards

As discussed in previous chapters, Adobe offers you several ways of creating additional artboards. These methods are as follows:

- You can add additional artboards by opening the **Artboards** window and selecting **New Artboard** from the **Options** submenu (see *Figure 9.7*).

- On the same **Artboards** window, you can also click the **New Artboard** button (+ sign) located at the bottom of this panel (see *Figure 9.7*):

Figure 9.7 – Adding artboards using the Artboards window

- On the **Properties** panel, select the **Edit Artboards** button. This will allow you to find the **New Artboard** button once again.

- On this same **Properties** panel, there is also the **Move artwork with Artboard** option. This is only visible after selecting the **Edit Artboards** button. When checked, you can drag a copy of any artboard by holding *Alt/Option* while dragging it to a new location (see *Figure 9.8*):

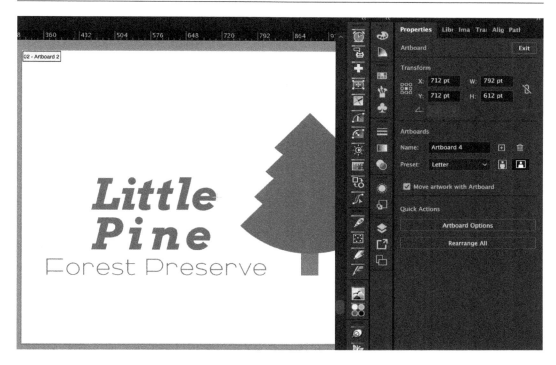

Figure 9.8 – Editing artboards using the Properties panel

- If **Move artwork with Artboard** is unchecked, you can now drag any current artboard to a new location to make a blank copy with the same dimensions as the original artboard.

- Also within the **Properties** panel, you will find the **Artboard Options** and **Rearrange All** buttons.

- Another way you can create new artboards or copy and edit current artboards is through the dedicated **Artboard** tool (*Shift + O*) located in the toolbar. This will engage the **Edit Artboards** workspace, just like selecting it from the **Properties** menu.

- In addition to having the **Edit Artboards** options available in the **Properties** panel, you can also find them in the *Options Bar* located directly below the menu bar at the top of the interface when in the **Essentials Classic** workspace.

In *Figure 9.9*, I have illustrated how you could add additional artboards by using the **Artboards** window, and how to rearrange them using the **Rearrange All** button. In this case, I chose **Grid by Row** for **Layout, Change to Left-to-Right Layout** for **Layout Order**, and **2** for **Columns**:

Figure 9.9 – Adding artboards using the Artboards window

Even more important than knowing how to add and arrange artboards is knowing how to customize them. In the next section, we will look at how to rename artboards, adjust their dimensions, and methods for sharing them.

Labeling, arrangement, and additional options

Just as there are multiple methods for creating and copying artboards, there are also several methods for adjusting current ones. To keep this simple, I will refer to the *Adding additional artboards* section of this chapter.

You have just learned in that section that there are many ways to work with artboards, but the two buttons we will need are the **Artboard Options** button and the **Rearrange All** button. The **Artboard Options** button brings you an options panel that allows you to choose from different preset sizes or enter a new customized size. This panel also offers you another place where you can rename the artboard. You will also be able to delete the artboard from this panel.

If you enter the **Edit Artboards** workspace, you can also click to select any artboard and then resize it by using its bounding box. If you decide that you need the additional options found within the **Artboard Options** panel, just double-click on the selected board. See *Figure 9.10* for a full list of options within the **Artboard Options** panel:

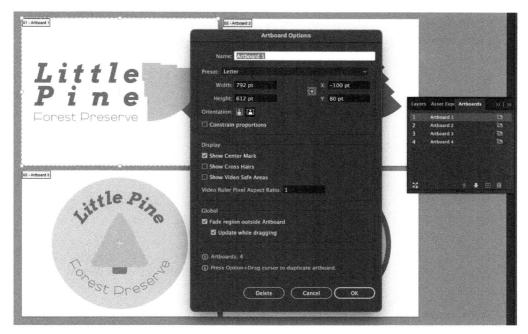

Figure 9.10 – The Artboard Options panel

The **Rearrange All** button gives you options for **Layout**, **Columns**, and **Spacing**.

If you deselect the **Move Artwork with Artboard** option, your boards will move while leaving your artwork where it is (see *Figure 9.11*):

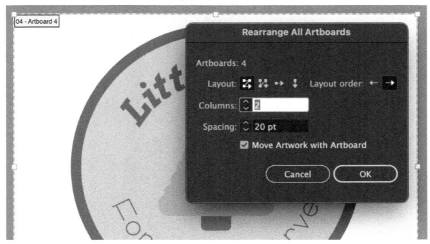

Figure 9.11 – The Rearrange All Artboards panel

In addition to these options for adjusting the layout appearance of your artboards, you can also choose to hide them entirely using **View** > **Hide Artboards** (*Shift + Ctrl/ Command + H*):

Figure 9.12 – Artboards visible (L) and hidden (R)

As you can see in *Figure 9.12*, the ability to hide the artboards brings about some unity to a collection of items. This may be beneficial when presenting your work directly from within Adobe Illustrator.

An additional option for either sharing your work to collaborate with a peer or for client review is the **Share Document** panel located at the top of the **graphical user interface (GUI)**. This requires that the document is first saved to your cloud account. You can then invite others to view and edit the file. You also have the option to allow them to make a copy, share the link with others, and comment. Click the gear icon at the top right to make these additional selections (see *Figure 9.13*):

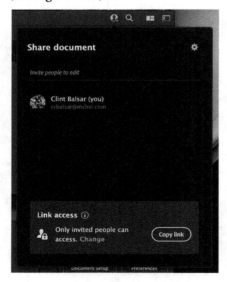

Figure 9.13 – Collaborate with peers or clients

Once invited, the recipient will receive an email notification, as you can see in *Figure 9.14*, to inform them that they have been invited to work with the file:

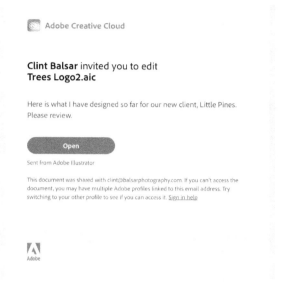

Figure 9.14 – Invitation email

The email includes a direct link to the web page of the associated file. When they click the **Open** button, it will take them to a page that looks similar to *Figure 9.15*:

Figure 9.15 – Creative Cloud file collaboration

As you can see in *Figure 9.15*, there is an area for comments (if allowed), file information, and history. It allows them to download the file, open it directly in Illustrator, or share it (if allowed). It also allows them to quickly scroll through the multiple artboards that are hosted within the file.

If you would rather not allow them to have access to edit the file, you could save it as a **Portable Document Format (PDF)** file. This will make your multiple-artboard file into a PDF file with multiple pages (see *Figure 9.16*):

Figure 9.16 – Artboards as multiple pages in a PDF

All these options allow you to reduce the number of project files. The ability to share a single file and still maintain the level of control you choose as the creator of the work is an invaluable tool. This is a huge step forward compared to the early days when a folder would be created for a project and then become bloated with individual files, additional versions, and resources. Cloud-based storage and multiple artboards have simplified the process a great deal.

Summary

As we discussed in *Chapter 8*, *Preparing Artwork for Presentation*, staying organized should be a goal you set each time you are developing your work. The use of artboards and their benefits will aid you in doing this. One of the greatest benefits of artboards that I reflect on and celebrate often is their ability to allow you to hold so much related content within a single file. The need to search for content related to a task has been greatly diminished by this tool. In addition, it bolsters your presentation quality when working with colleagues and clients. When collaborating with colleagues, labeling allows you to clarify what each artboard's purpose is and offers several related views that can be accessed through the **Artboards** panel. When working with clients, you can output the file to a PDF and each artboard is created as a page.

In the next chapter, we are going to get into the advanced capabilities of the **Layers** panel and focus on your ability to harness these capabilities through good development and management of your layers.

10
Advanced Layer Development and Organization

I believe a well-organized workflow is essential to the success of not only your work but also your relationships with others. This must be clear to you at this point in the book! You will likely be called to collaborate with others, and they will truly appreciate your ability to keep things organized and logical. Of course, the same can be said for clients, but to an even greater degree. This will be how they view you professionally. Your ability to master your files quickly and confidently both in their creation and, later, in their adaptation will present you as someone who can consistently get the job done.

In the previous chapter, we reviewed how artboards can offer great benefits when it comes to both organization and presentation. In this chapter, we will be taking a deeper dive into how layers can also be used to keep your documents organized.

To accomplish this, the chapter will be divided into the following main topics:

- Using the **Artboards** panel to your advantage
- Organizing layers with objects, documents, and artboards in mind
- Using groups, nested groups, and **Isolation Mode** for improved workflow

Technical requirements

To complete this chapter, you will need the following:

- Adobe Illustrator 2022 (version 26.0 or above).
- High-quality internet access may be required for some situations.

Using the Artboards panel to your advantage

In *Chapter 9*, *Utilizing Multiple Artboards*, we talked quite a bit about setting up artboards, as well as adding or removing them and editing them. Now, in this chapter, we are going to look closer at the panel for them. *Figure 10.1* shows a document that currently has four artboards:

Figure 10.1 – The Artboards panel

To add more artboards, you can click the **New Artboard** button (which looks like a plus sign inside a square) on the bottom row of the **Artboards** panel. To remove artboards, you can click the **Delete Artboard** button (which looks like a trashcan) in the lower-right corner of this panel. To adjust the stacking order of the artboards, you could click the **Move Up** or **Move Down** buttons, which are just to the left of the **New Artboard** button.

As was first presented in *Chapter 9, Utilizing Multiple Artboards*, in the far lower-left corner of the **Artboards** panel, you will find the **Rearrange All Artboards** button. Selecting this button will bring up the dialog box, which allows you to customize the way you would like to arrange the current artboards (see *Figure 10.2*):

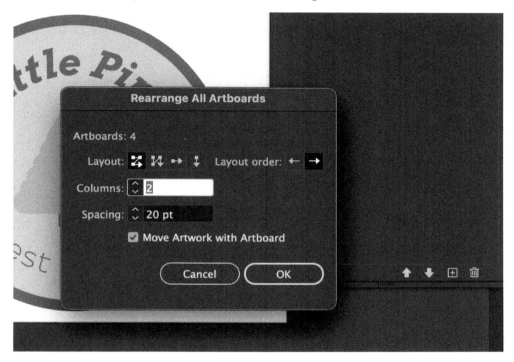

Figure 10.2 – The Rearrange All Artboards dialog box

Another benefit of artboards is the ability to navigate quickly from one to the other. While in the **Artboards** panel, click on any of the available artboards to make it active. Clicking it a second time will allow you to display the single artboard in the **Zoom to Fill** view. If you are looking to rename any of the artboards, you can double-click on its current name and then type in the replacement. If you would like to make edits to the artboard, you can also double-click on the **Artboard Options** button to the right of the name.

In addition, you can go to the top-right **Panel Options** button to find yet another way to make choices concerning artboards (see *Figure 10.3*):

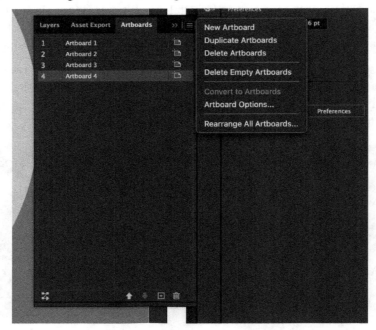

Figure 10.3 – Artboard Panel Options menu

You will be offered an additional choice by going to the **Panel Options** menu – **Delete Empty Artboards**. This will allow you to remove several artboards at one time if you did not apply any artwork to them.

You can adjust any artboard individually by selecting **Artboard Options…** from the **Panel Options** menu. This allows you to develop each board as a separate document with the correct dimensions. For example, you could have one artboard for a letter, one for an envelope, and one for a business card. When using this approach, you can easily compare the size of each. In addition, you will be able to easily print or export each item separately, if needed.

Next, let's talk about the methods you can employ when creating artwork with each of these artboards.

Organizing layers with objects, documents, and artboards in mind

As we discussed in *Using the Artboards panel to your advantage* section, multiple artboards allow you to have separate documents on each of them and even scale them to the exact size of the document. This allows you to host a variety of output media within one document, such as a business card, letterhead, and envelope design. One of the key benefits of preparing them on separate artboards is the ability to output each one as a separate asset when needed. To output artboards separately, you can use the **Export for Screens** command. Artboards also allow you to store and present multiple items or concepts within one file. If you do intend to collaborate or present an Illustrator file, then it will be very beneficial to keep everything logically organized.

In this section, we are going to look at how the development of the layers can maintain greater control of your work and result in less confusion and difficulty at a later point. Let's look at what we can do to combine shapes into object groups, place these groups into layers, and then position layers into the appropriate artboards.

In *Figure 10.4*, you can see that a single layer can hold the artwork for multiple artboards. This single layer has all the artwork, while in *Figure 10.5*, the image has the artwork hidden:

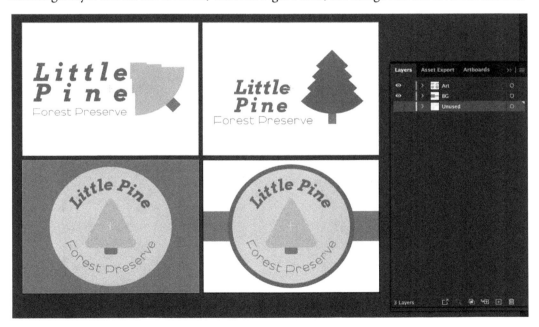

Figure 10.4 – A single layer may hold artwork for several artboards

Notice that with one click, the layer has been hidden on multiple artboards:

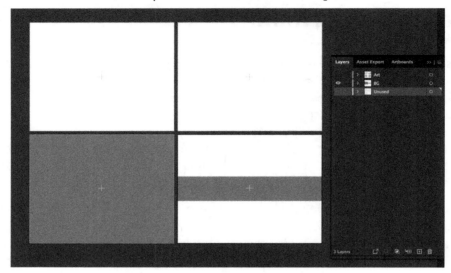

Figure 10.5 – All work becomes hidden when clicking the art layer

Although Illustrator allows you to have artwork from several different artboards within one layer, your work will be more organized if you produce your work in multiple layers. You can develop the layers based on the artboard that resides within them (see *Figure 10.6*):

Figure 10.6 – Creating a layer to hold artwork for each artboard

This will keep you more organized, with each artboard playing the part of a separate document with all its artwork on it.

In the example shown in *Figure 10.6*, each artboard's artwork was selected, made into a group (*Ctrl/Command + G*), and then moved to a newly created layer. Follow these steps if you are unfamiliar with this process:

1. Open the layer you have the current artwork in by clicking on the arrow to the left of the layer's thumbnail:

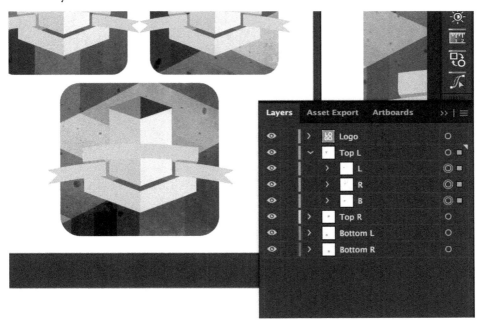

Figure 10.7 – Revealing sublayer items

2. Select all objects within a piece of artwork and choose **Object** > **Group** from the top menu (*Ctrl/Command + G*). In *Figure 10.8*, you will see that I have created a group for each shape. So, **Artboard 1** in the top-left corner has three groups:

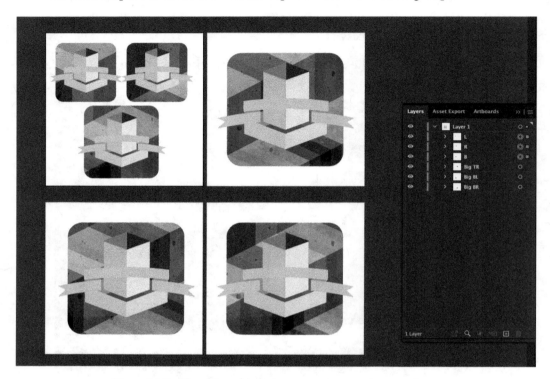

Figure 10.8 – Grouping objects to prepare to move to a new layer

3. *Shift + click* to select multiple groups, and then choose **Object** > **Group** from the top menu (*Ctrl/Command + G*) to collect them into one group (see *Figure 10.9*):

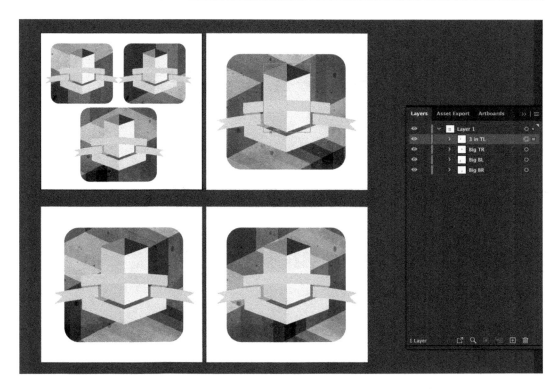

Figure 10.9 – Using a single group to migrate artwork to a new layer

The process of creating groups nested within a larger grouping is known as **sublayers**. Sublayers allow you to group details that are then grouped into the larger object.

4. Choose **Create New Layer** from the **Layers** panel.

5. Choose any of the sublayers from your previously created layer and drag them into the newly created one. The result of this action should appear like the **Top L** layer in *Figure 10.10*:

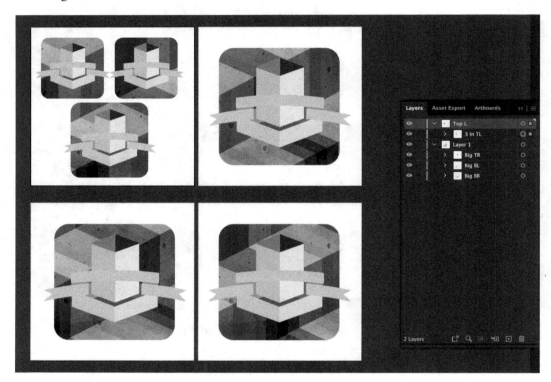

Figure 10.10 – Contents migrated into the new layer

6. Repeat step *4* for all remaining artwork. In this example, you would end up with four layers that each contains the artwork on their associated artboard (see *Figure 10.11*):

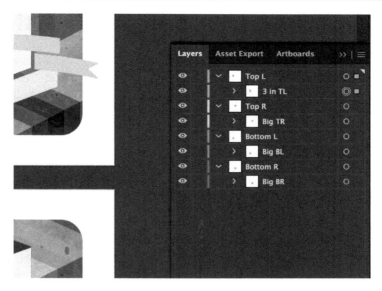

Figure 10.11 – A layer with sublayers created for each artboard

The flexibility of having layers with content that stays with the artboards, as well as layers that can hold content that crosses over several artboards, is powerful. Both techniques are valuable methods that you can use to your advantage.

Creating a layer that holds common items to apply to all artboards makes it easy to hide those same items all at once. For example, after creating a logo, you apply it to all content in every artboard, but then hide it using **Toggles Visibility** (the eye icon in the **Layers** panel). It creates a quick "before/after" for you to view the change.

Making sure your artwork is logically developed within layers by utilizing groups and subgroups may be a challenge at first. But, over time, if you develop the habits of layer organization, you will benefit from better time management and control of your work. Remember as well to manage your layers in relation to the artboards for even greater control. This will become more apparent when we discuss a popular editing technique in the next section.

Using groups, nested groups, and Isolation Mode for improved workflow

Any time you create an item consisting of several objects, those objects can then be selected and combined into one group. In *Figure 10.12*, you can see that I have selected both the green treetop shape and the brown tree trunk shape (hold *Shift* while clicking on each):

Figure 10.12 – Organizing artwork into a group

After selecting the objects you would like to be combined, navigate to **Object** > **Group** (*Ctrl/Command + G*) in the top menu. This will create a connected relationship between these shapes that makes it easy to perform further adjustments to them. Adjustments such as sizing and position can now be performed as though these objects are a single unit. When needed, the object can also be ungrouped to adjust individual components.

After creating a group, it will appear in the **Layers** panel as one item labeled **<Group>**. Just to the left of this label will be a small arrow pointing toward the right. Clicking on this arrow will allow it to rotate downward and reveal the objects included within the group. In this panel, you can now select any individual object within the group and then adjust it independently from the group it belongs to, as shown here:

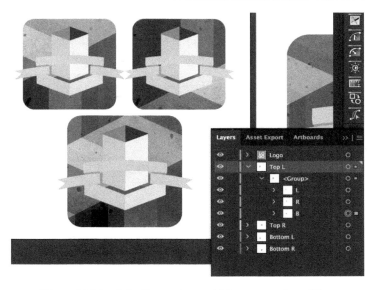

Figure 10.13 – Selecting an item within a group for editing

To select any object, go to the farthest right column within the **Layers** panel and then click to select it. In addition, you could change from the **Selection** Tool to the **Direct Selection** tool and then click on any object in the group. In *Figure 10.14*, you can see that the group has been opened to reveal the two objects that make it up; you can also see that one of them (the brown tree trunk) has been selected and then edited:

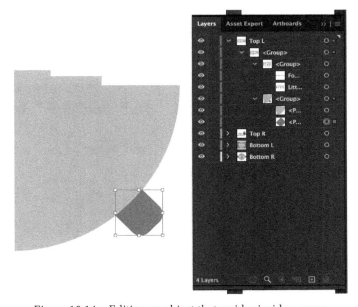

Figure 10.14 – Editing an object that resides inside a group

The benefits of groups can go even farther when you learn to use more complex nested groups. In *Figure 10.15*, the design for this layer is collected as a group and is made up of two additional groups that each have multiple objects collected within them:

Figure 10.15 – Nested groups

You can separate objects within any group using the **Ungroup** command. This can be done by first selecting the group you would like to separate, and then navigating to **Object > Ungroup** (*Shift + Ctrl/Command + G*) in the top menu.

If you have several objects collected within a group (or within several nested groups, such as *Figure 10.16*), it may be easier to simply go into **Isolation Mode,** make the intended changes, and then exit **Isolation Mode**:

Figure 10.16 – Objects collected in nested groups

Isolation Mode is a method by which you can quickly single out an element that is usually collected with a larger item, such as a group or nested group. To go into **Isolation Mode**, you need to move your cursor over the intended object inside your artboard and then double-click. This will adjust the view of your workspace and allow you to select each object or group inside the original group with one additional click. In *Figure 10.17*, I entered **Isolation Mode**, and then clicked on the group that holds the objects for the tree:

Figure 10.17 – Object selected using Isolation Mode

If I click one more time inside this isolated group, I will be able to select one of the two objects that make up this artwork. At this point, you can now edit the object, just like it was selected using the **Direct Selection** tool (which we first discussed in *Chapter 3, Developing and Organizing Objects*).

After making the intended adjustments, you can exit **Isolation Mode** by double-clicking outside the object or group currently in isolation. Other methods for getting out of **Isolation Mode** include going to the **Isolation Mode** bar located at the top of the document's window, right-clicking, and then choosing **Exit Isolation Mode** from the **Panel Options** menu. Another option is going to the **Layers** panel options menu and choosing **Exit Isolation Mode** (*Figure 10.18*):

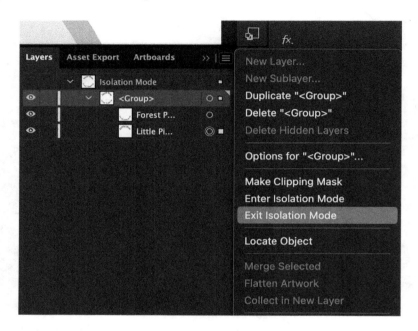

Figure 10.18 – Exiting Isolation Mode within the Layers panel options

I have often found that one of the quickest and simplest methods is to go up to the **Isolation Mode** bar and double-click on it.

If you are somewhat new to Illustrator, Isolation Mode may have just been that annoying thing that happens when you accidentally double-click on an item. If so, you have undoubtedly worked to exit it as quickly as possible. Now that you now know a bit more about Isolation Mode and its use within the structures of artboards, layers, groups, and subgroups, I'm confident that you will learn to make it one of your primary editing methods.

Summary

In this chapter, we have looked at the structure of the **Artboards** and **Layers** panels and discussed several methods for keeping your artwork in order, including working with **Isolation Mode**. As your work gains maturity and complexity, you will appreciate that you have acquired the habits that allow you to focus more of your time on the things that matter in your work and less on trying to find where part of your work is within the layer structure. Having a consistent workflow and structure will allow you to move into more and more elaborate designs over time. As your skills and confidence grow, so will the quality of your work. To this end, we took a deep dive into the benefits of collecting several objects into a group and the process of creating groups nested within a larger grouping (also known as sublayers).

In the next chapter, we will investigate how you can extend the power of Adobe Illustrator through plugins, and the options and benefits of the Illustrator app for the iPad.

11

Extending Illustrator Through Third-Party Tools and the iPad

By now, I hope you are getting more comfortable with your skills in Adobe Illustrator and gaining lots of valuable knowledge. As you continue to use what you have learned, you should see a marked improvement in your skill set. The best tool I can recommend to you is time. With time, your skills will continue to grow, and you will become more efficient using techniques, shortcuts, and additional tools.

Illustrator is already a great tool for creating vector artwork, but when we leverage the benefits that Illustrator offers us with additional tools, it gets even better. Whether it is through powerful third-party plugins, keyboard shortcuts, or even the iPad app, the benefits will be similar. Having these additional options will allow you to extend the capabilities of the software and break out from the rest that use Illustrator. This chapter is loaded with additional resources that can allow you to extend the capabilities of Illustrator.

To accomplish this, the chapter will be divided into the following main topics:

- Adobe Stock & Marketplace in your Creative Cloud app
- Powerful and creative plugins from Astute Graphics
- Cineware by Maxon for 3D objects
- Substance 3D in Illustrator
- Illustrator for iPad

Technical requirements

To complete this chapter, you will need the following:

- Adobe Illustrator 2022 (version 26.0 or above).
- High-quality internet access may be required for some situations.

Adobe Stock & Marketplace in your Creative Cloud app

As you continue to develop your skills and work to increase the quality of your designs, you will inevitably strive to gain a more professional appearance. Resources such as templates, stock images, and quality fonts will help you to move in that direction. Adobe has several resources (some included with your subscription and some requiring an additional fee) to allow you to expand your options. In the next sections, we will discuss each service offered by Adobe and what benefits you can obtain from it.

Adobe Stock

With the Adobe Stock subscription site (`https://stock.adobe.com`), you can look for a large variety of stock files to use within your own work. Available items include photos, illustrations, vectors, videos, audio, and 3D assets.

Adobe Stock is not included with your **Creative Cloud** subscription but can be utilized through an **Individuals**, **Teams**, or **Enterprise** monthly plan. You can also choose to pay as you go for each individual asset you need or prepay for a specific number of items (including 4K video, **Premium**, and extended licenses) using a product that Adobe calls a **Credit Pack**. You can purchase these packs with several different quantities of credits.

Figure 11.1 shows the website page for **Adobe Stock**, which includes a prominent search bar. To use the search bar, I recommend narrowing your search by first selecting a filter (located on the left side of the bar). By doing this, you will then only search for your intended media type instead of seeing results from the entire stock library:

Figure 11.1 – The Adobe Stock menu displays available media types

You could also utilize the menu at the top of the website to filter the results based on the content you are looking for. After finding an item you are considering, you can download a preview or save a preview to a **Creative Cloud library**. This will be a lower resolution and watermarked image. If you later decide to use this file, you can then upgrade it to a paid asset.

See *Figure 11.2* to view the **Download Preview**, **Save to Library**, **Open in App**, and **License** buttons:

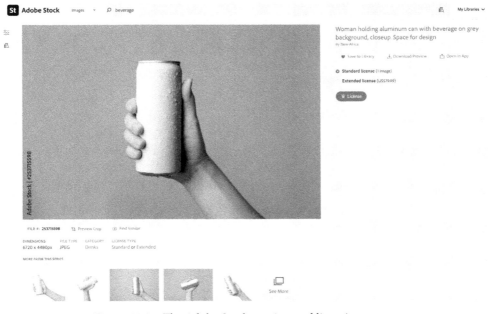

Figure 11.2 – The Adobe Stock preview and licensing screen

You can design a composition using the downloaded sample file, and then update it if you decide to share your work or offer it to a client.

When working with vector assets, you can choose to always download them as a vector file type, a JPEG file type, or to be asked each time. As shown in *Figure 11.3*, my preferred setting is to always download items as a vector file type, but of course, this is only for paid assets, as preview files will always be downloaded as lower resolution JPEG files:

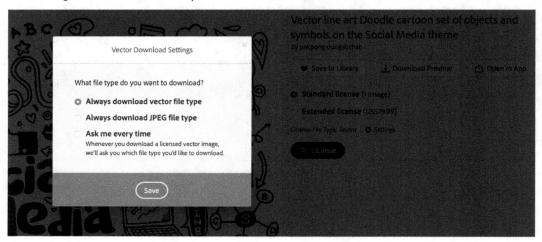

Figure 11.3 – Adobe Stock vector download settings

One of the greatest benefits of utilizing stock assets is the avoidance of copyright infringement. The fee is a small price to pay when your integrity is on the line. You will have peace of mind knowing that you have resourced additional assets to use with your work appropriately and legally. In addition to that, they will also raise the quality of your work by providing professional-quality items.

Adobe Fonts

Using the Adobe Fonts site (`https://fonts.adobe.com`), you can search for and add the perfect type to express the communication of your work. Adobe allows you access to a great collection of fonts that will help you set the tone or personality of your work.

Figure 11.4 shows the website page for **Adobe Fonts**, which includes a prominent search bar:

Figure 11.4 – Adobe Fonts displays available options in the submenu

The ability to have an extensive library of quality fonts is a great advantage to a creator, and Adobe now makes it available to anyone with a Creative Cloud subscription.

You can use the search bar, but I recommend narrowing your search by first selecting **Browse all fonts** (located directly below the bar). By doing this, you will then be taken to a web page (see *Figure 11.5*) that offers you a variety of filters on the left side of the page. The filters are divided into three categories – **TAGS**, **CLASSIFICATION**, and **PROPERTIES**:

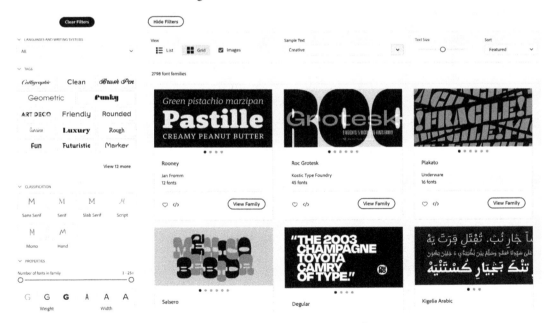

Figure 11.5 – Adobe Fonts user interface

You can search for the right font by using the website's filters menu.

Adobe Exchange

Using the Adobe Exchange site (`https://exchange.adobe.com/creativecloud`), you can expand the capabilities of your software by using third-party plugins. These plugins range from time-efficient workflow tools to design enhancement and template offerings.

Figure 11.6 shows the website page for **Adobe Exchange**, which includes a prominent search bar. To use the search bar, I recommend narrowing your search by first selecting a filter (located on the left side of the bar). By doing this, you will then only search for your intended media type instead of seeing the results from the entire stock library:

Figure 11.6 – Adobe Exchange is divided into two categories: Extensions and App Integrations

Substance 3D

The Substance 3D Community Assets site (`https://substance3d.adobe.com/community-assets`) allows you to find material assets that have been shared by others. It also allows you to share your assets with others.

Figure 11.7 shows the website page for **Substance 3D Community Assets**, which includes a prominent search bar at the top of the page. When using the search bar, you can narrow your search by first selecting a filter (located on the left side of the page). By doing this, you will then only search for what you are looking for instead of seeing results from the entire community library:

Figure 11.7 – Substance 3D interface

Creative Cloud libraries

After downloading assets from any of the previously mentioned sites, you can curate them in a Creative Cloud library (https://creativecloud.adobe.com/libraries) for easy access inside Illustrator (or any other Adobe title):

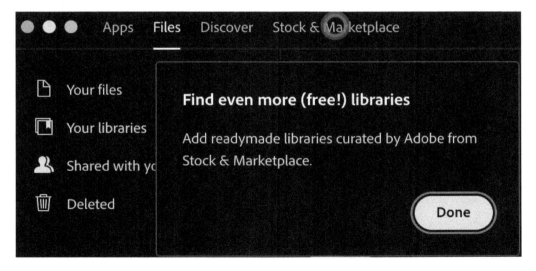

Figure 11.8 – Free curated libraries

You can also add additional files that you use to any Creative Cloud library. These additional files could include vector objects, images, type, brushes, patterns, templates, color swatches, and more.

Figure 11.9 shows the website page for curated **Creative Cloud libraries**:

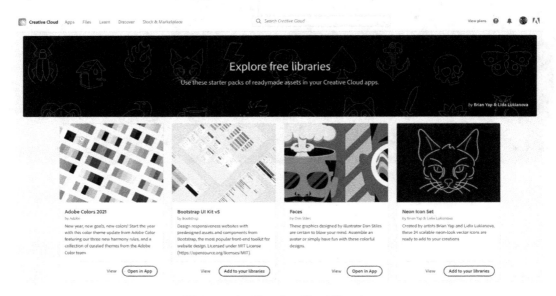

Figure 11.9 – Creative Cloud libraries

There is a huge number of resources available to you using these free shared libraries. Creative Cloud libraries also allow you to collect your own resources into a library, which can then be shared with and used by anyone who is collaborating with you on a project.

You can use the **Creative Cloud** app to find access to all these resources under the **Stock & Marketplace** tab, as shown here:

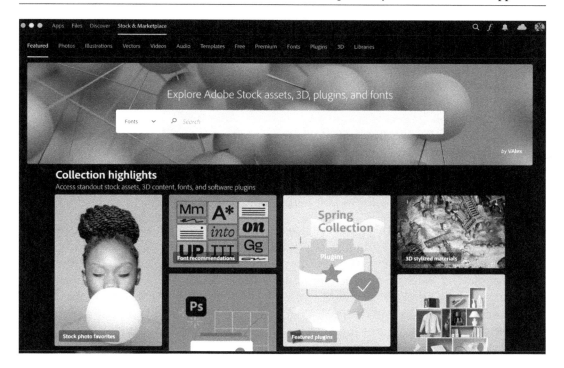

Figure 11.10 – The Stock & Marketplace section in the Creative Cloud panel

Adobe has done an excellent job of curating content that you may find useful for your workflow and hosting all of it directly in the app. This is also where you can find additional tutorials and update all your licensed Adobe software.

In this section, I have introduced you to each of these resources individually, so that I can highlight them one at a time. *Figure 11.10* shows the **Creative Cloud** app's menu, which includes the **Apps**, **Files**, **Discover**, and **Stock & Marketplace** tabs. By selecting the **Stock & Marketplace** tab, you will find the links to every resource we have just gone through.

By using the benefits of the Creative Cloud Libraries, you will be able to bring in resources that you have created and saved or those you have acquired from any of the Stock & Marketplace options. This will up your efficiency game quite a bit, as you won't be wasting valuable time searching for items such as branding, type, and colors if you have prepared a project library first.

Next, we will look at even more ways to extend Illustrator's abilities and further increase your efficiency with the software.

Powerful and creative plugins from Astute Graphics

As you get more and more comfortable with the tools already offered to you within Illustrator, the inevitable need to ask even more from it will occur. It might be based on a desired look or even a specific printing requirement. Although Adobe has done an excellent job developing Illustrator through the years, there are still areas where a third-party developer can devote a great deal of time and resources to bring even more out of the software. A third-party developer has the benefit of having a more limited focus than Adobe, since Adobe is looking at how everything works within Illustrator and even how it interacts with the additional software titles offered by their company.

There are many **plugins** out there, but the one that I go to most often and has the most opportunities to extend the capabilities of what Illustrator can do for me is the **Astute Graphics (AG)** collection (`https://astutegraphics.com`). The collection offers 21 different plugins. Four are considered free and included with a free 7-day trial (Astute Buddy, AutoSaviour, DirectPrefs, and MirrorMe), while the other 17 are all included in a yearly subscription. In addition to the 21 plugins, the subscription will give you updates on current plugins, new plugins when they become available, texture packs, brush sets, graphic styles, and more. They also have lots of training videos and knowledgeable customer support.

Let's look at what each AG plugin does and the opportunities it offers to you for improvement. This is not intended to be an extensive list of their functions and capabilities, but rather a brief overview of each, so you can get a feel for what each one is intended for.

Variety

The apps offered by AG perform a variety of tasks, and it might help to explain each app. In the following list, with each app, the available options are listed as described directly from the AG website, with a few additional thoughts from me:

- **AstuteBuddy** (free): **Keyboard shortcut panel** – Keeping this panel open as you utilize a different AG plugin will show pertinent keyboard shortcuts for that plugin.

- **Autosaviour** (free): **Autosave, backups, and reminders** – This tool helps by autosaving at a predetermined period and shows you the last time the document was saved. It can also be set up as a reminder for you to save if that is what you prefer.

- **ColliderScribe**: **Precise alignment and selection** – This tool helps you to build relationships between shapes.

- **DirectPrefs** (free): **Nudge distance, angles, and guides** – This plugin will allow you to customize your preferences for nudge distance, common angles, and guides.

- **DynamicSketch**: **Intuitive vector sketching** – When using this tool, you will gain a more natural sketch-like feeling.

- **FindReplace**: **Instantly locate and modify objects** – This plugin allows you to search out many similar items, and then make batch replacements.

- **InkFlow**: **Drawing and lettering in one tool** – This plugin intuitively draws smooth strokes with a more natural gestural feeling.

- **InkQuest**: **Pre-press controls and checks** – This is a great plugin to prepare your work for final approval and print.

- **InkScribe**: **Precise path creation** – This plugin assists you in adjusting your vector paths more logically. Simply select the path between two anchors and bend it to the arc that you want.

- **MirrorMe** (free): **Instant symmetry** – This is a great tool for both bilateral and radial symmetrical art creation.

- **Phantasm**: **Instant color control and halftone** – With this plugin, you would swear you are in Photoshop with the ability to adjust hues, opacity, values, and more. In addition, it can also process a vast variety of halftone patterns.

- **Randomino**: **Randomize in Illustrator** – This really allows you to creatively experiment within your work, as it will give you extensive control while developing a very randomized appearance.

- **Rasterino**: **Image crop and editing** – This plugin allows you to trim transparency areas from any raster image placed within an Illustrator file. You will also be able to edit, crop, and link any raster-based file.

- **Reform**: **Shape and manipulate** – With this plugin, you can reform the path of objects by selecting the path and pulling it to a new location. This doesn't require the level of knowledge and skill that adjusting paths using the anchors and direction handles requires.

- **Stipplism**: **Live stipple effect** – This plugin adds dense random dot patterns to represent the appearance of a stipple drawing.

- **Stylism**: **Live effects made easy** – This plugin allows for several effects that can be adjusted with live previews. Items such as drop shadows, feather, blur, and inner or outer glow can be explored without a final commitment.

- **SubScribe**: **Create accurate artwork** – This plugin straightens or connects paths for greater accuracy of your work. It also locks and unlocks objects directly.

- **Texturino: Textures and opacity brush** – This applies raster-based texture images to any shape. Through the user interface, you can adjust scale, opacity, and blend mode.

- **VectorFirstAid: Clean up vector documents** – This plugin makes an inventory of your document and discovers inaccurate and unnecessary elements and then removes them with a single click.

- **VectorScribe: Editing, shapes, corners, and measure** – This plugin allows intelligent adjustments to your vectors. It cleverly removes extra anchors, and edits and extends paths.

- **WidthScribe: Variable stroke width effects** – With a pull from your mouse, you can expand or contract the width of any path.

Methods of use

The apps that you acquire through a subscription offer a wide range of capabilities. Whether your work is graphic or illustrative in nature, the AG plugins can be a huge benefit to your workflow. They can also be very helpful to the final output quality. To illustrate the point, I have shared one of AG's online lessons for creating infographics and what my results were in *Figure 11.11*:

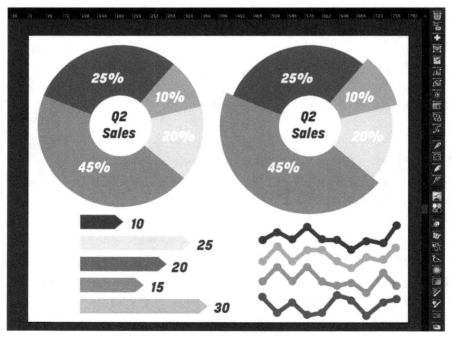

Figure 11.11 – Infographics created using the Dynamic Shapes plugin

Astute offers tutorials for each plugin they offer, as well as several quality lessons for completed projects. The AG infographics lesson that I used can be found at `https://astutegraphics.com/learn/tutorial/how-to-create-infographic-elements-with-vectorscribe-in-illustrator`. This lesson allows you to become better acquainted with the capabilities of the Dynamic Shapes plugin while creating completed infographic elements.

In contrast, here is an illustration I created using a variety of AG's plugins (*Figure 11.12*):

Figure 11.12 – Illustration created using several AG plugins

Many additional industries make use of third-party extensions to Illustrator, but graphic design, fashion design, and illustration are the three most prominent trades that make use of them.

I created the bug illustration to illustrate the use of AG plugins. The illustration benefited from the use of many AG plugins, but really highlights the benefits of the Stipplism, Texturino, and Stylism plugins.

> **Additional Information**
>
> I have included a link that was shared on the AG website, which highlights an extensive collection of additional third-party plugins: `https://astutegraphics.com/learn/tutorial/third-party-illustrator-plugins`.

To take a deeper dive into what the AG plugins are capable of, I would recommend researching a few of these artists, who make excellent use of these tools:

- Aaron Draplin: `http://www.draplin.com`

- Von Glitschka: `https://www.glitschkastudios.com`

- Dave Clayton: `https://heshootshedraws.com`

To reach a professional level of production within Illustrator, practice your skills and techniques as often as possible, and even extend the capabilities of Illustrator when needed to get to what you truly intended. The most important thing to remember is to be creative! Tools can help you create your concepts but being able to conceptualize something intriguing and unique is the "secret sauce" that you must bring to any creativity tool you are using.

In the next section, we are going to look at an enhancement to Illustrator that allows you to map your artwork directly onto complex prebuilt 3D models.

Cineware by Maxon for 3D objects

Maxon is the creator of Cinema 4D and has been in the 3D workspace for many years. You can find out more about them and the products they offer at `https://www.maxon.net/en/`. They have now partnered with TurboSquid (a stock 3D store owned by Shutterstock) to offer a plugin that will work with selected 3D models offered under the Cineware brand. These objects can work directly within Illustrator and allow a simple method for 2D artwork mapping onto an assortment of items represented by 3D models, such as packaging and apparel. You can find out more about the plugin and access a download link at `https://www.maxon.net/en/cineware-illustrator`. After downloading and installing the plugin, you should visit `https://www.turbosquid.com/cineware/browse` to find a selection of both free and paid 3D models to download and use the Cineware plugin with.

The plugin has a custom workspace that you need to set up before beginning to import any of the TurboSquid 3D templates. Go to the top menu and choose **Window** > **Workspace** > **Cineware 3D**. This will give you three panels that are part of the Cineware plugin package. These three panels are **Scene Structure**, **Attributes**, and **Materials** (see *Figure 11.13*):

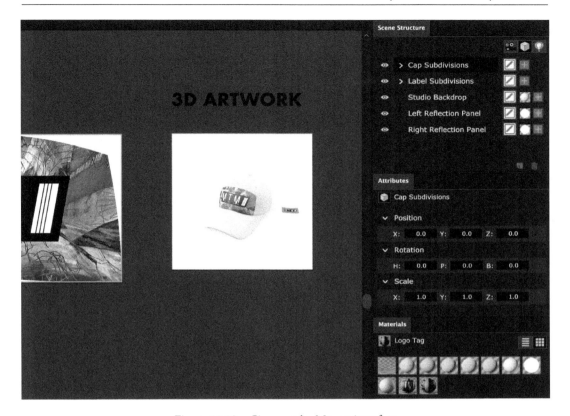

Figure 11.13 – Cineware by Maxon interface

From the **Scene Structure** window, you can open the subdivision that you intend to adjust (in the case of this baseball cap, it would be the **Logo Space** subdivision), and then select that subdivision's material (represented by a small sphere icon). This will open the properties for the material in the **Attributes** panel. Inside this panel, you should find a **Texture** attribute that may already contain the image for that panel. You can apply your adjusted texture simply by dragging it to your selected attribute and dropping it on top of the previous instance.

Adobe has continued to add valuable 3D options and improvements to Illustrator. In the next section, we will look at how you can use the built-in offerings to create 3D objects, apply and edit materials, and render the final output.

Substance 3D in Illustrator

Through the acquisition of Substance 3D, Adobe has begun to integrate this product into the Creative Cloud subscription in interesting ways. You can now benefit from some of the free materials offered within the updated 3D user interface within Illustrator. Adobe has decided to offer the complete Substance 3D package as an additional subscription, but you can still do quite a bit with the free materials, as it gives you the ability to add your own materials or even wrap a graphic. You can also add lighting for a final touch before rendering your image.

For a thorough introduction of the latest 3D options offered in Illustrator, including the integration of Substance 3D, check out this section of the Illustrator User Guide: `https://helpx.adobe.com/illustrator/using/create-3d-graphics.html?trackingid=YB1TGLWS&mv=in-product&mv2=ai`.

This guide offers simple instructions divided into three steps:

- **Create a vector art**
- **Apply 3D effects to the artwork**
- **Render the artwork**

The first and the third steps are part of most Illustrator workflows and are quick and simple, but with 3D, we should take a moment to discuss step *2*.

As you will see in the second step, **Apply 3D effects to the artwork**, the **3D and Materials** panel (**Effect > 3D and Materials**) has three sections that offer options for transitioning your vector artwork to 3D. The three sections are as follows:

- **Object**
- **Materials**
- **Lighting**

The **Object** section will allow you to select which type of 3D you would like by selecting one of the **3D Type** buttons at the top of the panel.

Illustrator can develop vector artwork into **Plane**, **Extrude**, **Revolve**, and **Inflate** types of 3D. In *Figure 11.14*, you can see that I have simulated a cork by first drawing a simple rectangular vector shape, then using the **Revolve** type to develop it into 3D:

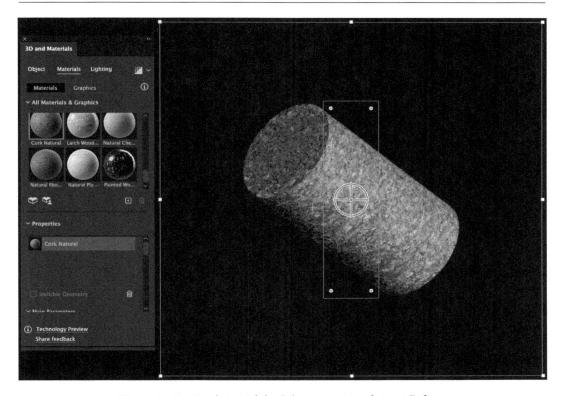

Figure 11.14 – Applying Adobe Substance materials to a 3D form

Using the **Materials** section, I have applied the **Cork Natural** material offered by the free **Adobe Substance Materials** library that comes with Illustrator. In the third section, you can make final lighting decisions before rendering the scene out as an image file. To render the image, use the button on the top right of the **3D and Materials** panel and choose your intended render results.

Adobe has been making steady improvements to the 3D panel and it has become a valuable element of the Illustrator offerings. Another area of steady development is the iPad version of Adobe Illustrator. In the next section, we will look at what you can do with this tablet version of Illustrator.

Illustrator for iPad

Adobe has been working hard at developing a natural workflow between the desktop and iPad versions of Illustrator. Using the new cloud AI format (.aic), your work can be saved to the cloud and then travel between the two Illustrator apps. At the time of writing this book, I would not recommend this for all situations, as there are still many things that the iPad version cannot do. In fact, I have found that it makes some destructive changes to files occasionally. The good news is that cloud files always have a revisions list and can be brought back to a previous state.

To get acquainted with what you can do with Illustrator for iPad, follow along as I build a quick card design that can later be opened in Illustrator on the desktop for further enhancements. Before beginning the project, it is worth noting that a quick way of learning about different tools and methods of working with them in this version of Illustrator is to explore the *help* icon (third from the right at the top of the GUI). This offers a very thorough library of tool hints, shortcuts, and tutorials.

As you can see in the screenshots I have included in *Figure 11.15*, starting a file is still similar in this version of the software, but due to the touchscreen capabilities and the Apple Pencil option, there are several iPad-specific techniques to get familiar with:

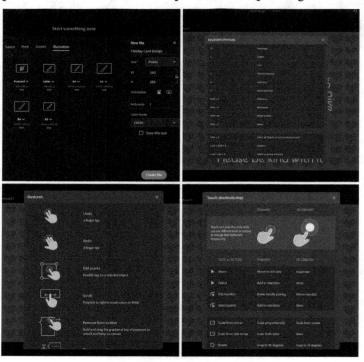

Figure 11.15 – Various menus found in the help section of Illustrator for iPad

The following activity is intended to give you something to create that will allow you to get more familiar with the iPad version of Illustrator and consider what you might want to use it for. I have found that it allows you to get an artwork started, even when you are away from your desktop (or laptop), and then you can choose to continue working on it with the desktop version. In this activity, we are creating a holiday card design that will require building objects with simple shapes, the use of text, and the use of patterns:

1.   To begin this design, choose **Rectangle Tool** and create a background shape that encompasses the entire artboard. In *Figure 11.6*, you will see that I created a rectangle with a violet fill. Also, note the location of the **Layers** panel in this figure, as we will be utilizing it often throughout this project. The **Properties** panel is located directly below the **Layers** panel, and we will also make use of that quite a bit:

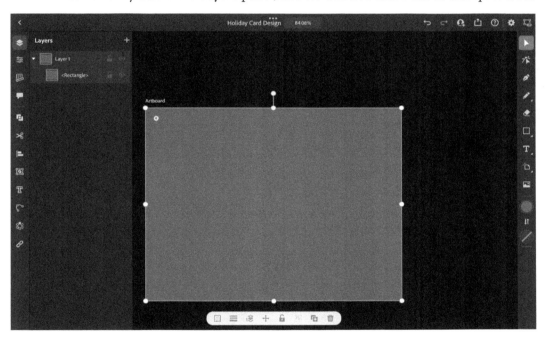

Figure 11.16 – Creating a background using Rectangle Tool

We are now going to go through a few steps to develop a heart shape. Granted, there are several options for creating such a shape, but I have found this to be an interesting way to show how to make use of the tools within Illustrator and Illustrator for iPad.

2. Start by creating a rectangle that is approximately twice as tall as it is wide using the **Live Corners** widget to maximize the corner radius of all four corners. This will result in a pill shape, as shown here:

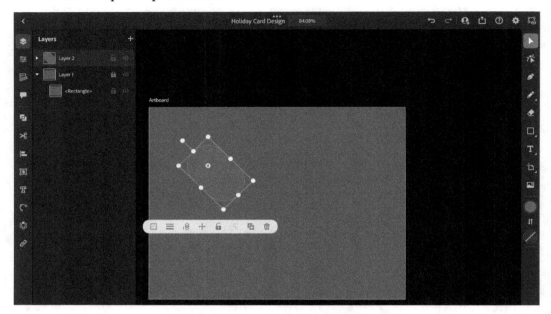

Figure 11.17 – Using the Live Corners widget to create a pill shape

Use the **Rotate** handle on the outside of the object's bounding box or use **Properties > Transform** to rotate this shape to 45˚. When using the **Rotate Tool**, you can also hold the center of the **Touch Shortcut** tool (gray circle in the lower left of the GUI in *Figure 11.16*) to get Illustrator to snap the rotation to 45˚. The center of the **Touch Shortcut** tool is known as the primary shortcut. If you were to pull the **Touch Shortcut** tool to the outside of the circle, a secondary shortcut would occur. In this case, the selected object would then snap to 10˚.

Now that we have a single pill shape angled to 45°, we need to make the bottom point. Using the **Direct Selection** tool, we will select one of the lower anchors and then pull the **Live Corners** widget outward to develop a sharp corner. Next, we will need to make a duplicate using the **Common Actions** bar located directly below the selected object (See *Figure 11.17*). The icon located second from the right will create a duplicate, and you can then go to the taskbar located on the left of the GUI to select **Flip > Vertical**.

3. We will now move the mirrored copy over to create the second top curve of the heart shape. To be sure the alignment is correct, I would suggest changing the **View** mode to the **Outline** mode, which can be found at the top-right corner of the GUI:

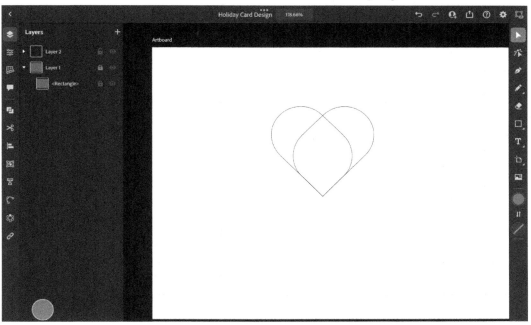

Figure 11.18 – Check alignment of overlapping shapes in Outline mode

4. After making sure the alignment is correct, we will select both overlapping halves with the **Selection tool**. You can click to select the first shape, and then hold the **Touch Shortcut** tool while clicking on the second shape to add it to the selection. In the area to the left that we now know as the taskbar, you will find the **Combine Shapes** panel, which includes the **Shape builder** tool. We will use this tool to draw a line across the two objects. This will create a new shape resulting from their combination (see *Figure 11.19*). Also, note that you will want to hit **Done** near the top of the workspace to close the **Shape builder** tool:

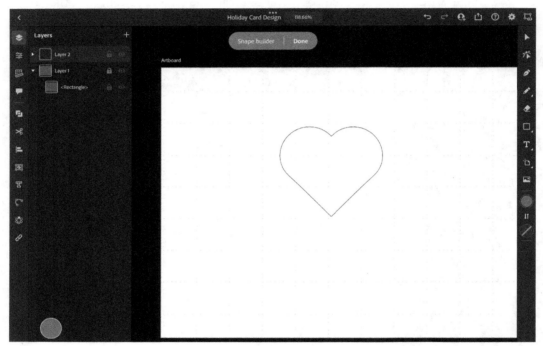

Figure 11.19 – Resulting shape after using the Shape builder tool

5. Let's again go up to the top-right corner where we can select the **View** mode and change it back to **Preview**. We are now going to start to develop an illustration of a 3D box using **Rectangle Tool**. Draw out a rectangle that has a similar shape to the one you can see in *Figure 11.20*. This will not need exact dimensions. Just remember that it is being used to cover one side of the heart shape, so make sure it overlaps one side of the heart and extends beyond the outside of the shape a little bit as well, as shown here:

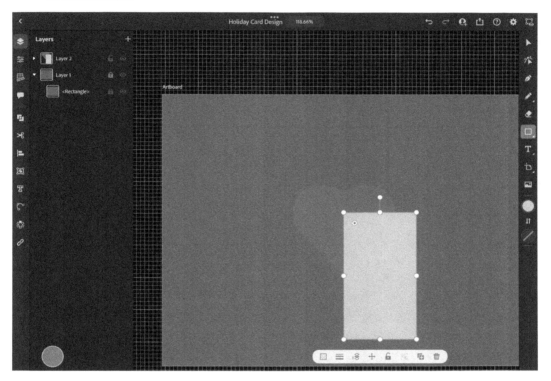

Figure 11.20 – Creating a box panel using Rectangle Tool

6. Use the **Direct Selection** tool to select the top-right corner of the bounding box. Hold down the **Touch Shortcut** tool and add the bottom-right corner to your selection. Click and drag to move both anchor points upward to simulate dimension to this side of the box:

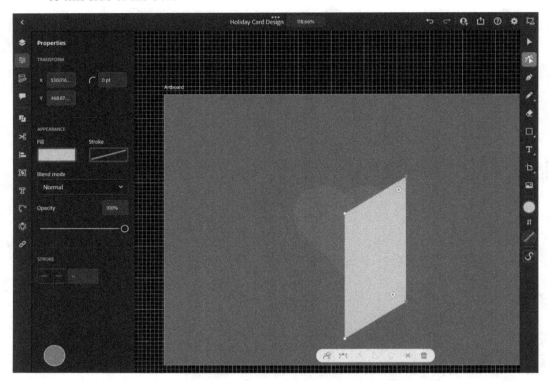

Figure 11.21 – Using the Direct Selection tool to select multiple corners

Holding the **Touch Shortcut** tool will allow you to constrain the movement if you would like it to only move upward, but not outward.

7. Now, we will use the **Selection** tool to select this first side of the box and create a
    duplicate using the **Common Actions** bar located directly below the object. As was
    pointed out in step *2*, you can duplicate a selection using the icon that is second
    from the right. To adjust this shape to reflect the first side, we will again go over to
    the taskbar on the left and choose **Flip** > **Vertical**. You may recall that this is the
    same method we used in creating the second side of the heart shape in step *2*. To
    complete this step, we will need to move the flipped copy over using the **Touch
    Shortcut** tool to constrain the movement:

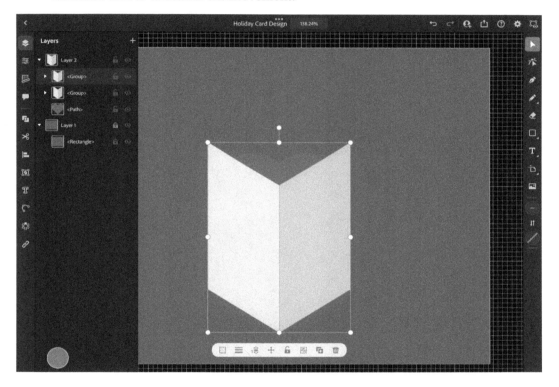

Figure 11.22 – Utilizing Duplicate and Flip to create a second box panel

Try to align the center edges so that there is neither overlap nor a gap. Adjust the
color of this second side of the box so that it is a tint of the first side. This will
represent the side of the box that is receiving more light.

Now that we have the front of the box created, we will be using it to create the back
of the box.

8. Let's first select both sides using the **Selection** tool. Then, open the **Object** panel located in the taskbar and select the **Group** command. With the newly created group still selected, use the **Common Actions** bar to create a duplicate, and then, use the **Layers** panel located in the top left of the GUI (or you can also use the icon that is third from the left in the **Common Actions** bar) to move this duplicate below the heart shape. To give this box the feeling of perspective, we need to flip this bottom group. Select the group, go to the taskbar, and select **Flip > Horizontal** (see *Figure 12.23*):

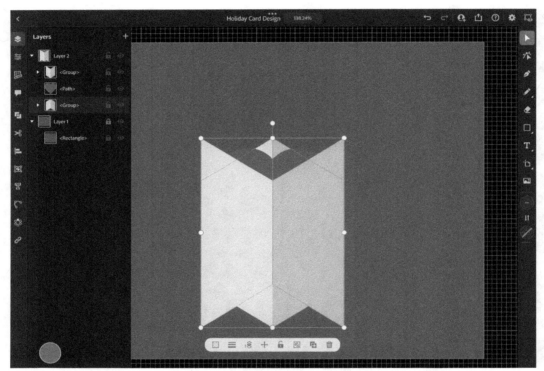

Figure 11.23 – Duplicating a group and arranging the layer order

You will need to move this group up to meet the edges of the front of the box. Hold down the **Touch Shortcut** tool to constrain the movement. I would recommend you lock the other objects using the lock icon in the **Layers** panel, as they may inadvertently get selected as you are trying to work with the back of the box grouping.

9. To make this represent an open box, we will next need to customize the colors so that it appears that a single light source is illuminating it. A great technique for getting to individual items with a group is to first select the group and then double-click on the object within the group you would like to select, as shown here:

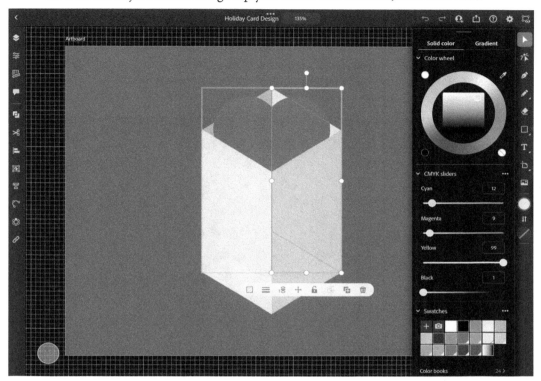

Figure 11.24 – Adjusting an object within a group after double-clicking with the Selection tool

You can then open the **Fill** attributes and use the eyedropper to the top right of the color wheel to select the opposing color from the front of the box. For example, if you have selected the back-right box shape, you will fill it with the same color as the front left. If you selected the back left, then you will initially fill it with the same color as the front right. I like to adjust the back left to have a slightly darker shade, as it is an inside panel so would be receiving less light. I would recommend setting the color space to be CMYK and then moving the black slider to the right by a small amount.

10. Be sure to move the heart to a level that gives the appearance that it is partially in the box, and we can then consider this icon complete. Remember that you may need to unlock the layer, group, or object. In Illustrator for iPad, if you see a lock icon on the top left of a bounding box, you can click this icon to unlock the item. While you still have the heart selected, you can create a duplicate that will be used to develop a pattern for the background. After creating the duplicate, I would recommend moving it into your background layer. Then, using the **Selection** tool, let's select the new heart and move it off to a side of the artwork. Click and drag one of the corner anchors while holding the **Touch Shortcut** tool to constrain the scale as you reduce the size in preparation for making a pattern from it:

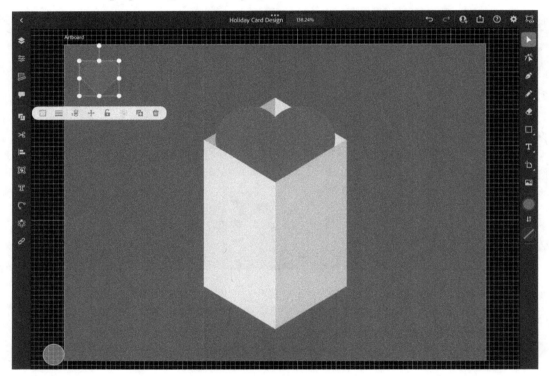

Figure 11.25 – Scaling a duplicate of the heart shape

11. While the smaller heart is still selected, go to the taskbar and choose **Repeat** > **Grid** to begin creating a repeating pattern that can fill the background. Changing the bounding box size using a corner will change the size of the objects within the pattern, so make that decision before moving to the next step. Now, move the entire bounding box to the top-left corner of your work, and then drag the handles to the right and downward to add more rows and columns. This will decide the boundaries for this background pattern. To complete the customization of the grid, drag the sliders to decide on the vertical and horizontal spacing between the repeating objects:

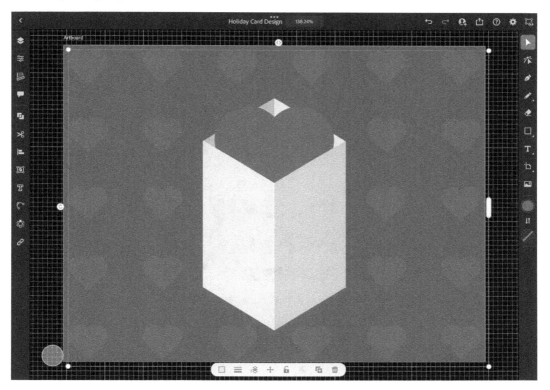

Figure 11.26 – Developing a pattern utilizing the Repeat tool in Grid mode

12. You can do further customization by utilizing the **Grid Repeat** properties located in the **Properties** panel. In this example, I used **Brick by Row** and **Flip Column > Flip Vertical**:

Figure 11.27 – Customizing the pattern using Repeat options as well as Blending Mode and Opacity

13. To develop the message for this card, let's prepare to use **Type on Path**. Create a new layer (using the + icon in the **Layers** panel). In this new layer, we will use the **Ellipse** tool to draw out a circle that is slightly bigger than our main art but leaves space for the type that will be outside of it (look for a diagonal Smart Guide to be sure it is a perfect circle). Go to the taskbar and use the **Align** tool to make sure things are falling into the center of your design. You might notice that the main artwork has been grouped at this point. This will need to be done before selecting the **Horizontal Align Center** and **Vertical Align Center** buttons. After checking on the heart in a box artwork, do the same centering for the circle behind it. Now, using the **Common Actions** tool, create a duplicate of the circle. This extra circle will be used to assign the **Type on Path** text.

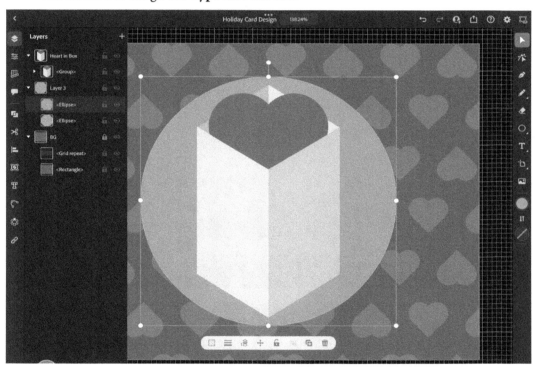

Figure 11.28 – Creating a circle and then aligning and duplicating in preparation for the Type on Path tool

14. To create the text, we will select **Type Tool** and then draw out a textbox to contain the phrase we would like the card to say. Feel free to use what I have used or get creative! Use the **Properties** panel to choose an appropriate font and customize the text attributes, such as size and color. Select the new text and add the top circle shape, and then from the taskbar, choose **Type** > **Type on Path**:

Figure 11.29 – Applying text to a path and utilizing Type on Path options

Your text should now be wrapping around the ellipse shape but might not be located along the path the way you would like. Find the left alignment bracket (single line) and the right alignment bracket (double line) and place them at the beginning and end of the intended arc for the text. Then, under the **Properties** panel, choose **Paragraph** > **Align Center** to distribute it evenly between the two guides.

15. As a final element, I chose to apply a secondary phrase in a textbox just below the artwork. This allows for a secondary phrase as well as giving a balance to the overall design. Again, I centered the textbox and then chose **Paragraph** > **Align Center**. Now that you have reached the end of this project, review the appearance and make any design adjustments you feel might improve the design. For example, I decided I wanted to try out some different colors and ultimately chose to recolor the large circle to coordinate more with the rest of the color palette:

Figure 11.30 – Finalizing the design with additional text and organizing layers in the Layers panel

Although the organization of layers is not required, I would highly recommend once again that you consider it for your own benefit. You are likely to use Adobe Illustrator for iPad as a method for starting your work but will usually revisit it in Illustrator on your desktop, and the naming and placement of layers will make it easier for you to jump in where you left off.

It is also going to be highly appreciated if you are sharing the document. While still in Illustrator for iPad, you can quickly export as PNG and then share the file as a composite layout (comp) for approval (see *Figure 11.31*):

Figure 11.31 – Design as created in Adobe Illustrator for iPad

Illustrator for iPad works well with the Apple Pencil and thus becomes a great alternative to a drawing tablet (such as a Wacom Intuos Pro tablet). Having both an Apple iPad and a drawing tablet adds additional benefits to your workflow, but either one will help you get the job done with more control than a mouse generally can.

Although both the desktop and iPad versions have some similarities, they still differ quite a bit when it comes to their features and capabilities. Storing your files in the cloud allows you to easily move back and forth between the desktop and iPad versions while keeping the files synced (see *Figure 11.32*):

Figure 11.32 – GUIs on desktop (left) and iPad (right)

One of the best benefits of Illustrator for iPad is the ability to continue working on your vector-based artwork when a desktop or even a laptop is not convenient. You can also work offline and then update it when you once again have a connection to the internet.

Summary

The additional options for Illustrator allow you to customize important tasks that Illustrator might not have been specifically designed for or is currently capable of doing.

As I have said in previous chapters, it is not only about gaining skills but also becoming more efficient with them. The plugins and extensions I have highlighted in this chapter will save you valuable time as well as improve the quality of your work.

In this chapter, you have learned that using the Creative Cloud app will connect you to a large volume of resources. More specifically, when using the **Stock & Marketplace** tab, you will gain access to a variety of websites that Adobe is offering specifically to enrich your artwork through additional resources and knowledge. We have also looked at the AG line of plugins to extend the capabilities currently offered in Illustrator. We finished this chapter by exploring the use of 3D options through both the third-party plugin known as Cineware by Maxon, as well as the built-in capabilities of Illustrator and the inclusion of Substance 3D.

In the next (and final) chapter, we will try to pull together all that you have learned to this point and put it into action. We will discuss why efficiency is such a valuable asset. Most creatives understand the necessity to gain knowledge and skills (myself included) but fall short in increasing their ability to use them efficiently. We will look at methods for gaining and practicing several time-management behaviors, such as keyboard shortcuts and focusing apps.

12
Illustrator Mastery – Advanced Techniques and Shortcuts

In this final chapter, we are going to take a closer look at some possible output projects you may be faced with and some associated skills that would help you with each challenge. We will look at some workflow methods, as well as highlight several options available to you to enhance your Illustrator artwork.

In this chapter, we are going to discuss the benefits of drawing in both **Preview** and **Outline** modes of view, as well as developing a variety of patterns, textures, and type options.

To accomplish this, the chapter will be divided into the following main topics:

- Logo development using the Outline view and Shape Builder
- Mood board development using Pattern and Repeat for background fills
- Comp card with advanced typography settings
- Illustration with texture fills

Technical requirements

To complete this chapter, you will need the following:

- Adobe Illustrator 2022 (version 26.0 or above).
- High-quality internet access may be required for some situations.

Logo development using the Outline view and Shape Builder

When designing the branding for an organization, I would recommend that your conceptual phase for the design be initiated on paper. You will have the freedom to sketch out a large number of options very quickly. This should allow you to develop multiple options much quicker than trying to illustrate them on a computer.

As an example of this technique, I have sketched several possible logos for a fictional video production team called Better Than One Productions, consisting of two co-owners. I want to relay the importance of them being a successful team that works well together, while each offering their individual talents and ideas to the outcome. After creating several possibilities, I chose to build a draft of the lower-right sketch (see *Figure 12.1*):

Figure 12.1 – Planning phase starting on paper

The lower-right sketch is made up of several simple shapes, which should make it an ideal logo to use for presenting the development using the two strategies of this section. Use one of your own sketches and follow along as I show you the steps I recommend when developing your artwork using the **Outline** view and the **Shape Builder** tool:

1. After planning out a few possible designs, it is time to import them into an Illustrator file and use them to continue your ideation process. There are several ways by which you can take your sketch into Illustrator, so you should choose whichever works best for you and your workflow. You could scan it using either a traditional scanner or an app scanner on your phone or tablet. Just be sure to keep the scan resolution low. A high-resolution image is not required for this technique, because you are just using it as a guide. You can also take a photo of it using any digital camera, including your phone or tablet. I like to use this method and then AirDrop it to my MacBook Pro. If you do not use Apple products, another technique you can use is to move the image through a cloud storage service, such as Google Drive or Dropbox. Save it using the app for the service of your choice, and then retrieve it from this service on your workstation. If you use a digital camera, you can either connect it via a USB cable or read its card through a card reader plugged into the workstation. No matter which method you choose, you will need to have the file on your workstation before adding it to the Illustrator file.

2. Your next step is to create a new file. I would recommend considering your final output requirements, but just remember that vector-based art is not resolution-dependent, so it can be adjusted later.

3. To begin creating from your sketches, go to **File** > **Place** (*Shift + Command/Ctrl + P*) and select the sketch file you have brought into your workstation. Select the **Options** button in the bottom-left corner of the **Place** dialog box, as shown here:

Figure 12.2 – Placing an image

This will allow you to see additional options that you can choose at the time of placement. If you would like to keep your file size smaller, you can choose to link this placed file rather than have it embedded, which is the default. You can also choose to identify this image as a template here, but it will then be placed at full resolution. I prefer to make the image a template after first scaling the image to the artboard.

4. You should now notice that your sketch is represented as a thumbnail version of the original image and attached to your cursor. The is known as a **Placegun** and it can hold multiple images for placement. If you were to hit *Return/Enter*, it would place the image at full resolution, but I recommend that you click and drag to scale the image to your desired size. When it appears correctly, you can release the mouse.

5. To prepare this image to be used as a template, go to the contextual menu within the **Layers** menu and choose **Template**. You can also choose **Options for "Layer1"…** and then add a checkmark for **Template**. You should now see a new icon (a rectangular frame) for the visibility of the layer, as well as the layer's lock icon (see *Figure 12.3*):

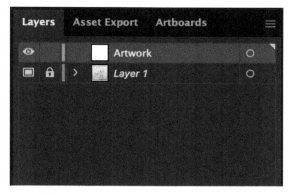

Figure 12.3 – Preparing a Template layer

6. You will now need to create an additional layer (or layers) to develop your artwork on top of the now dimmed **Template** layer (see *Figure 12.3*). The template layer continues to use the name **Layer 1** but can be changed to `Template` (or any other name) if you so desire.

7. After beginning to develop your artwork over the template, you will soon discover that there will be times when you would prefer to see the vector paths you are creating rather than the additional properties being applied to those vectors, such as fills, gradients, textures, and strokes. On such an occasion, I would highly recommend the use of toggling between the **Preview** and **Outline** views (*Ctrl/Command + Y*).

In *Figure 12.4*, in the left image, you can see the early stages of drawing out some of the base shapes, while in the right image, I have used the **Outline** mode to consider the vector locations and relationships:

Figure 12.4 – Preview and Outline views

The **Outline** mode will allow you to see through objects to understand where the underlying vectors are in relation to the shape you are creating.

8. To create more advanced shapes, you can use the **Shape Builder** tool to add to or subtract from a shape. First, draw a shape using any available vector path tool, and then draw an additional shape that overlaps it.

In *Figure 12.5*, I have created a shape to represent a face and then an additional shape to represent the hair:

Figure 12.5 – Stages using the Shape Builder tool: Two overlapping shapes in Preview mode (top left) and Outline mode (top right). Using the Shape Builder tool to remove a portion by holding down Alt/Option (bottom left) and the result of removing that portion (bottom right)

Note that the top vector paths of the shape for the hair have extended beyond the top of the face.

To use the **Shape Builder** tool, you will need to hold *Shift* and click any shapes that are intended to be used. Then, select the **Shape Builder** tool and either draw through the shapes to add them together, or hold *Alt/Option* and click to remove a shape resulting from the overlap of two or more of the original shapes.

9. Now, using the **Selection** tool, make sure the current shapes are deselected. Then, select each individual resulting shape and add custom attributes, such as **Fill** and **Stroke**, to complete the object (see *Figure 12.6* for an example):

Figure 12.6 – Shape Builder results with Fill attributes applied

By using the **Shape Builder** tool, you can be confident that items, such as the hair in this example, fit perfectly onto other items, such as the face. Many complex forms can be created by simply adding to or subtracting from the original form.

After developing your custom logo draft, it will be necessary to present it to others. Whether it is colleagues you need to collaborate with or clients you need to present it to, you will want to share it in a professional manner so it represents the best it can be. In the next section, we'll look at a couple of methods to present it clearly and professionally.

Mood board development using Pattern and Repeat for background fills

For this section, we will work through a possible scenario that you might come across. When preparing a **brand guide** or **mood board** for a current or potential client, it is often nice to reiterate the appearance of their brand as a watermarked pattern in the background. If this process does not directly relate to what you do within Illustrator, it may still connect to an unrelated situation where you could still benefit from the building of patterns, such as a character's clothing, or wrapping paper on a package.

We will review several methods and options for creating patterns within Illustrator and look at what **Astute Graphics** can offer you for pattern creation through their **MirrorMe** plugin.

Pattern

You can easily create patterns of several types within Illustrator. You can build a pattern out of any part of your art, but you must remember to select it before using any of the pattern tools. The original pattern builder is located by going to the top menu and choosing **Object** > **Pattern** > **Make**. This will present you with the interface to begin making choices for how you would like your pattern to draw out (see *Figure 12.7*):

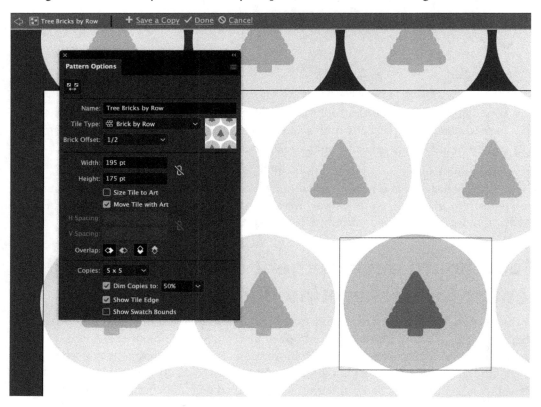

Figure 12.7 – The Pattern Options interface

The five tile types you can choose from are **Grid**, **Brick by Row**, **Brick by Column**, **Hex by Column**, and **Hex by Row**. Using the **Pattern Options** panel, you will be able to select the **Pattern Tile Tool** and customize the tile spacing used when creating the pattern. After selecting the tool, you will notice that the tile edge will become an active selection with corner and side anchors. Enlarge the frame beyond the artwork to create greater spacing, or reduce the frame into the artwork to create overlapped spacing.

After choosing all the options for the pattern, you can add a name for the pattern, select the **Save a Copy** command on the **Isolation Mode bar**, and then select **Done** to close the Isolation Pattern. Now, you can find this new pattern in your fills to be used within any shape. In *Figure 12.8*, I have applied a recently created **Brick by Row** pattern to a rectangular shape:

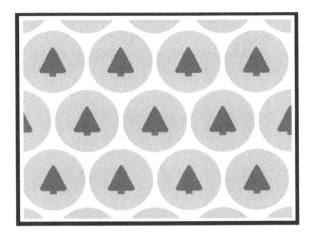

Figure 12.8 – Result of using the Brick by Row pattern

With this technique, the saved pattern has become a pattern **Fill** attribute and can be adjusted using the **Scale** and **Rotate** tools. Double-clicking on either of these two tools will bring up the options for them, where you can then select **Transform Patterns** to adjust the pattern rather than the object.

If you do choose to overlap the pattern, be sure to try out the four available methods, **Left in Front**, **Right in Front**, **Top in Front**, and **Bottom in Front**, to control the orientation of the overlap.

At times, you might find that you want to develop a pattern that is not created into a **Fill** attribute, but rather as an attribute that can accept live editing within an object. Illustrator has added the **Repeat** tool for just such a situation, and I think you will find it very helpful.

Repeat

Illustrator has another option for creating patterns, which you may find allows you even more choices in the pattern design. This next technique is a newer pattern tool for Illustrator and is divided into three options, which we will discuss next.

Repeat – Radial

The first one we are going to look at is the **Radial** method for repeating an object. To initiate this pattern, you must first select the item or items you would like to be included within this pattern. Next, go to the top menu and select **Object > Repeat > Radial**. You should now see the result of that action.

These newer options offer a wonderful **widget** that allows for direct editing of the pattern from within the bounding box of the pattern. Using *Figure 12.9* for visual reference, let's look at the elements of the widget that allow you to customize the radial pattern:

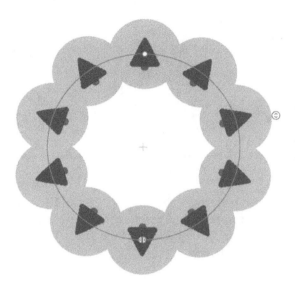

Figure 12.9 – Result from Repeat – Radial

On the far right is an icon that, when used, allows you to increase (by sliding it upward) or decrease (by sliding it downward) the number of instances repeated in the pattern. This is known as the **Instances** control. On the top is an icon (a small white dot) that, when used, allows you to increase (by sliding outward) or decrease (by sliding inward) the radius of the radial pattern. In addition, this icon allows you to adjust the angle of the pattern by rotating it left or right. This is known as the **Circle** control.

In *Figure 12.10*, what appears to be an icon at the bottom of this widget is actually a pair of icons. This is known as the **Splitter** control and is used to reduce the number of instances along the radius of the radial pattern. By grabbing the right side of the icon and sliding it counterclockwise, it will reduce the pattern in that direction, while grabbing the left side of the icon and sliding it clockwise will reduce the pattern in the opposite direction:

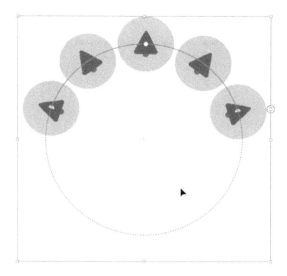

Figure 12.10 – Splitter used to reduce instances

You can also drag from any corner of the bounding box to increase or decrease the size of the pattern.

The benefit of the **Repeat** method of creating a radial pattern is the ability to quickly edit it from within its bounding box. The widget options allow you to work quickly and edit the pattern in several ways without the concern of a permanent change.

Repeat – Grid

If a radial design is not what you need but you would benefit more from a grid pattern, **Repeat** can do that as well and still offer similar options for quickly updating the pattern.

The second option we are going to look at is the **Grid** method for repeating an object. It also offers a widget that can be used to customize the size and quantity of the resulting pattern. To initiate this pattern, you must first select the item or items you would like to be included within this pattern. Next, go to the top menu and select **Object** > **Repeat** > **Grid**. You should now see the result of that action.

You can now use the widget to adjust this pattern. Pulling the handle on the bottom will allow you to add additional rows to the pattern while pulling the handle to the right will allow you to add additional columns to the pattern.

Dragging the top or left sliders allows you to increase or decrease the spacing in the grid. All four of these widget icons are visible in *Figure 12.11*:

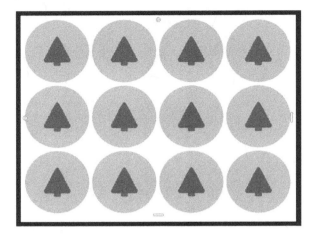

Figure 12.11 – Result from Repeat – Grid

You can also drag from any corner of the bounding box to increase or decrease the size of the pattern.

Again, the benefit of the **Repeat** method of creating patterns is the ability to quickly edit it from within its bounding box. The widget options allow you to make changes with immediate results.

Repeat – Mirror

After creating the first half of an artwork you intend to have a symmetrical pattern in, Illustrator can offer several options on how it reflects a copy. For this, we have the third option in the repeat family – the **Mirror** option.

To initiate this pattern, you must first select the item or items you would like to be included within this pattern. Next, go to the top menu and select **Object** > **Repeat** > **Mirror**. You should now see the result of this action.

You can now use the widget to adjust this pattern. Dragging the center widget handle will allow you to control the spacing between the original and mirrored artwork. This line, which hosts this widget handle, is known as the **symmetry axis** (see *Figure 12.12*):

Figure 12.12 – Result from Repeat – Mirror

Using the handles above or below the artwork on the symmetry axis will allow you to rotate the mirrored artwork. If you would like Illustrator to rotate both the original and mirrored versions of the artwork, just use the **Rotate** methods (such as grabbing the **Rotate** icon outside the bounding box or engaging the **Rotate** tool) on the original artwork, and the mirrored version will rotate in the opposite direction.

To complete the edit, just double-click outside the selection. Both halves of the artwork will now be grouped together and will move as one object. Double-click the artwork if you would like to make further adjustments to the **Mirror** pattern.

An added benefit to all three of the **Repeat** commands is the **Repeat Options** panel (**Object** > **Repeat** > **Options**). This allows you to have greater control after the pattern is initially made. The panel will vary depending on which **Repeat** technique is used within the selected object.

Another option, which is not available inside Illustrator but can be added as a paid plugin, is MirrorMe, which is available from Astute Graphics. This tool is more flexible than the **Mirror** Repeat method within Illustrator. Let's take a closer look at the additional options it offers.

Astute Graphics – MirrorMe

An additional option that can be used, but does not come from Adobe, is the Astute Graphics **MirrorMe** plugin. You can choose to select the art you would like to build the pattern from and then hit the farthest left button, **Symmetry axes will affect selected artwork only,** or just select the button to the right of the previously mentioned button, **Symmetry axes will affect entire layer and be retained**, to affect all content within the current layer.

You can then move the segment that will be used to create the mirrored elements. Left-click on the artboard to confirm the choice. The angle and number of segments will determine the result of the final mirrored creation. Both the angle and segments can be adjusted by either using the **MirrorMe** panel or double-clicking on the artboard. This will bring up **Edit Symmetry Axes** (see *Figure 12.13*):

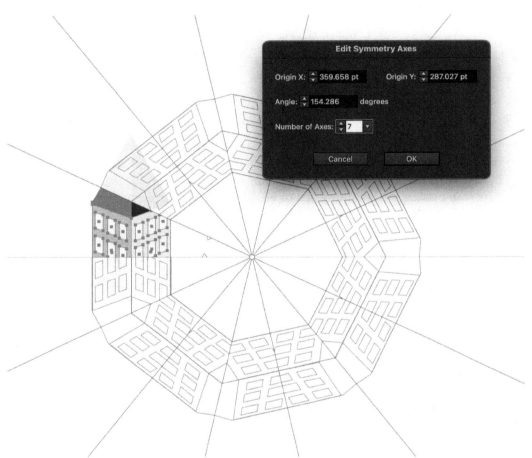

Figure 12.13 – User interface from Astute Graphics' MirrorMe

After editing the mirrored selection, you can accept the new pattern by using one of the buttons located at the top left of the workspace, as follows:

- The first choice, **Apply to Selection**, is a good choice if you would like to create a new object out of the pattern and then have the MirrorMe interface appear again for consideration of an additional MirrorMe pattern.

- The second choice, **Apply to Layer (persistent)**, is a good choice if you would prefer to have the MirrorMe interface stay persistent on the layer. This allows you to move the resulting pattern to create a new pattern.

- The third choice, **Cancel**, will allow you to abort the current effort.

By using patterns built from **MirrorMe** or any of the other options mentioned earlier, a background watermark can be placed on a guide board and then reduced to create a subtle repetition of the new brandmark (see *Figure 12.14*):

Figure 12.14 – Completed brand guide using a pattern fill-in background

The patterns you create for a **brand guide** could appear in several places and/or situations within the board, such as a patterned textile or a patterned area of a package design. You might not always utilize a pattern in presenting the brand to your client, but if needed, you should feel confident that a pattern can be created and applied to add that extra bit of authenticity to the proposal.

In the next section, we will be looking at another popular task that requires more advanced typography skills than the previous project. It can easily be adjusted to any print project you might have, such as promotional postcards or even greeting cards. The focus will be on how to customize the text within the document.

Comp card with advanced typography settings

So far, we have been concentrating more heavily on the drawing and compositional aspects of creating Illustrator artwork, but projects often require a good amount of text design, as well.

Let's go over a few items you should be aware of when developing any project that requires a considerable amount of text, such as a talent comp card. This publication will require several entries based on the talent's statistics. Although the tradition of the comp card is rooted in the modeling industry, variations can be made for actors, musicians, and others.

To compose all this content, I would recommend using several different textboxes and using the alignment tools to arrange them. This will allow you to rearrange your composition with more flexibility, as you are looking for the best layout. Here are some additional suggestions for using different text layouts in your project:

- **Selection of a font**: Readability should be a key factor when selecting a font for professional documents. Remember the adage "*form follows function*" when considering any typography. It can help your design through its aesthetics but must also be successful functionally.

- **Font adjustments**: Refer to *Chapter 7, Powerful Typography Options in Adobe Illustrator*, for a review of the methods for working with **Font** attributes. The **Character** panel will allow you to make necessary adjustments to scale, kerning, leading, and more, as shown here:

Figure 12.15 – Character panel options

For this comp card, I have selected a clean **sans serif** font for the model's name and purposely selected a font (in this case, **Nort**) that can have several variations. I used the **Extralight** style for titles, and the **Bold** style to designate specific content for each title.

To further adjust the text inside a multiline textbox as we have in *Figure 12.16*, I like to select the textbox with the **Selection** tool (*V*), and then scroll over the available options within the **Character** panel to see a preview. I adjusted the leading between these two lines of content with a final choice of **21 pt**:

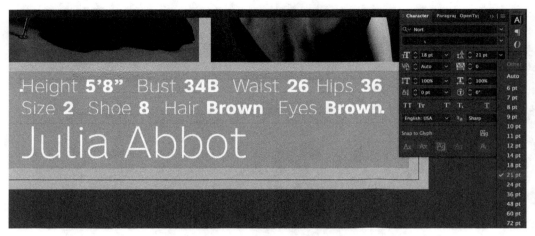

Figure 12.16 – Adjusting leading

Alignment can also play a big part in cleanly designing with type. In this example, I chose to align these two lines by using **Justify all lines** from inside the **Paragraph** panel, because they were already very close to matching up. I also chose to align the model's name with the content above it with **Horizontal Align Left** from inside the **Align** panel by using **Align with Key Object**:

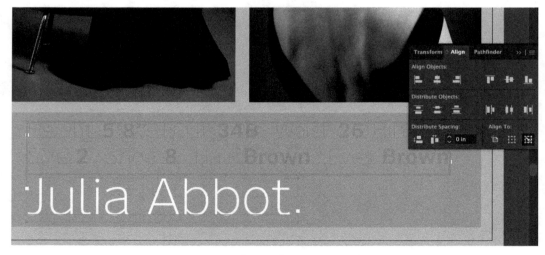

Figure 12.17 – Align with Key Object selection

When using this technique, you must be sure to select (by clicking on) the object that you want to use as the key object. This will ensure that the key object will not move, and the additional item(s) will be adjusted to complete the intended alignment.

- **Paragraph adjustments**: The **Paragraph** panel will allow you to choose the method of justification for any paragraph of text. With the **Paragraph** panel, you can make choices such as **Align Left**, **Align Center**, or **Align Right**.

  You could also choose **Force Justify**, so that text aligns to both the right and left of the textbox. These are several options based on how you would like to end the last line:

  a. **Justify with last line aligned left**

  b. **Justify with last line aligned center**

  c. **Justify with last line aligned right**

  d. **Justify all lines**

  This panel also allows for additional options such as **Left indent** and **Right indent** to move the text inside the actual textbox. You can also choose to only apply this option to the first line with **First-line left indent selection**.

  To give some room at the top and/or bottom of a paragraph, the options of **Space before paragraph** and **Space after paragraph** are available to you. All these options allow you to adjust the amount by adjusting the point's value in the interface.

  The last option on the **Paragraph** panel is a checkbox that allows you to decide whether you would like Illustrator to hyphenate words or not.

As this chapter is intended to encourage you to explore advanced techniques and shortcuts, I wanted to add a couple of tips that I use each time I work with text. When you want to adjust the spacing between two characters (which is also known as kerning), you can use the **Character** panel, but I prefer an alternative option. Using the **Type** tool (*T*), place the insertion cursor between the two characters, and then hold *Alt/Option* and use the *left* and *right* arrow keys on the lower right of your keyboard to loosen or tighten the kern between these characters.

In the next section, let's look at how textures can help your work and how some advanced techniques in their use can bring your work to a new level of professionalism.

Illustration with texture fills

When you began using Adobe Illustrator, you might have been focused on achieving smoothly-curved shapes drawn with vector lines and anchors that could be enhanced with simple attributes such as **Fill** and **Stroke**. Then, as you acquired more skills and learned more about the software, you likely began expanding the attributes of the objects you created. A stroke was changed to a more unique brush appearance and widened. A fill was given a gradient instead of a solid. All these attributes will enhance your artwork, but if you would prefer vector artwork with the look of a more painterly illustration, the application of textures is what you need.

I think this may be one of the areas of Illustrator in which users are most confused, especially regarding its capabilities. It is fairly apparent how to apply a pattern swatch, but additional tasks, such as how to use it with other attributes and how to customize its scale and appearance, are often hard to learn. In this section, I hope to enlighten you on some of the advanced techniques and shortcuts I utilize to create unique patterns within objects.

Applying texture with a pattern swatch

To create a shape with a texture, select the object and then create a copy using *Ctrl/Command + C*. Then, choose **Paste in Front** by using *Ctrl/Command + F*. On this new sublayer, select a texture swatch by using the **Swatch** contextual menu. Select **Open Swatch Library** > **Patterns** > **Basic Graphics** > **Basic Graphics_Textures**. After selecting the texture of your choice, you can scale it by double-clicking the **Scale** tool, and then choosing any increase or reduction in scale. Be sure to have the **Transform Patterns** option checked, leaving all other options unchecked:

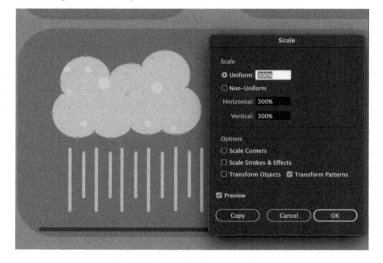

Figure 12.18 – Scaling a texture

An alternate method that can be used to transform any texture is to hold the *tilde* key while adjusting a bounding box.

While the texture is still selected, go to **Recolor Artwork**, and then choose **Advanced Options…** to access additional choices. Next, select **Color Reduction Options** (located between the **Preset** and **Colors** selectors) and then uncheck **Black** under the **Preserve** category. This will allow you to recolor the texture (see *Figure 12.19*):

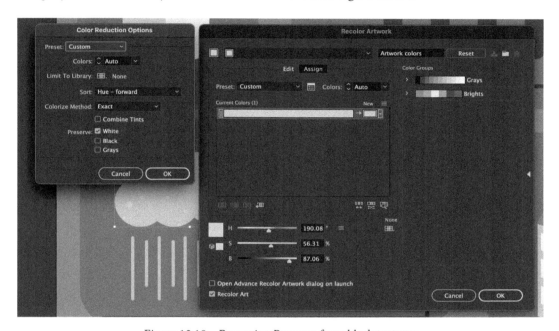

Figure 12.19 – Removing Preserve from black textures

Using this technique, you will be able to add a variety of rich textures as a sublayer on top of a solid or gradient fill.

Applying texture using a raster-based image

Using a very similar technique, you can apply any raster-based image (photograph, etc.) to express a texture. Move any image directly below the copied object and then select both the image and the copied sublayer. After that, right-click, and choose **Make Clipping Mask**. This will clip the image into the object. Using **Transparency** (**Window > Transparency**), you can blend the image to blend with the original object below. In *Figure 12.20*, you can see that I chose to use **Hard Light** for **Blending Mode**, and **65%** for **Opacity**:

Figure 12.20 – Applying transparency to a raster-based image

An additional method you could use with similar results would be to **image-trace** the raster-based image. This technique will create a vector-based version of the original and allow you to customize the appearance of the resulting tracing. The **Image Trace** panel (**Window > Image Trace**) will allow you to make several choices on the vector output (see *Figure 12.21*):

Figure 12.21 – Image Trace panel options

The **Image Trace** panel allows you to use presets or customize the output using a variety of sliders. You can choose whether the resulting tracing will be **Color**, **Grayscale**, or **Black and White** under the **Mode** option. The **Colors** slider will then allow you to choose how many colors or levels of gray will be used to represent the original image.

Applying texture using Astute Graphics – Texturino

The addition of a third-party plugin offers many additional options for applying textures as well as an extensive library of textures to choose from:

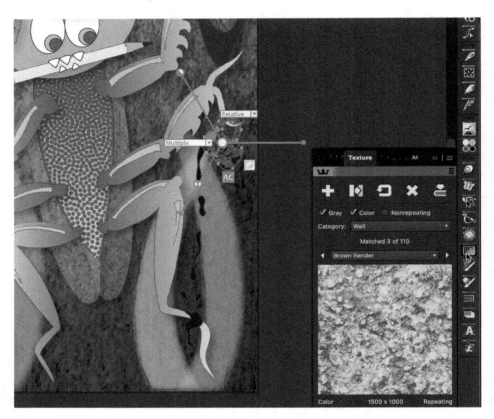

Figure 12.22 – The Astute Graphics Texturino palette and adjustment widget

The adjustment widget will allow you to make changes to **Scale**, **Center Offset**, **Opacity**, and **Blend Mode** of the applied texture image.

No matter which method (or methods) you choose to use to apply textures, I hope you find that doing so adds an additional layer of richness to your otherwise smooth and clean artwork. Of course, there are often times when textures are not needed within your vector work, but having the knowledge and power to use them when needed allows you to extend your skill set for anything that might come your way.

Summary

In this final chapter, we have reviewed Illustrator's more advanced techniques and tools for efficiency. The use of keyboard shortcuts will accelerate your workflow a great deal, and, over time, will aid you in advancing your work and getting challenging designs completed in less time.

I hope you have enjoyed learning with me. No matter what your Illustrator level was when you began reading this book, I hope I have offered some content that has taken your knowledge and/or skills past that point now. Illustrator becomes an extension of your hand as far as creativity goes so, like traditional artistic media, the longer you work with it, the stronger your skills grow. Each artist finds their own approach to using their media. So, each Illustrator artist finds their own methods for its use. I wish you the best as you continue on your journey to becoming more comfortable with this tool and allowing it to free your creative spirit.

Index

W

Z

Packt.com

Subscribe to our online digital library for full access to over 7,000 books and videos, as well as industry leading tools to help you plan your personal development and advance your career. For more information, please visit our website.

Why subscribe?

- Spend less time learning and more time coding with practical eBooks and Videos from over 4,000 industry professionals

- Improve your learning with Skill Plans built especially for you

- Get a free eBook or video every month

- Fully searchable for easy access to vital information

- Copy and paste, print, and bookmark content

Did you know that Packt offers eBook versions of every book published, with PDF and ePub files available? You can upgrade to the eBook version at packt.com and as a print book customer, you are entitled to a discount on the eBook copy. Get in touch with us at customercare@packtpub.com for more details.

At www.packt.com, you can also read a collection of free technical articles, sign up for a range of free newsletters, and receive exclusive discounts and offers on Packt books and eBooks.

Other Books You May Enjoy

If you enjoyed this book, you may be interested in these other books by Packt:

Up and Running with Affinity Designer

Kevin House

ISBN: 978-1-80107-906-8

- Explore the interface and unique UX characteristics of Affinity Designer
- Discover features that allow you to manipulate and transform objects
- Apply color, shading, and effects to create unique compositions
- Employ layers to organize and simplify complex projects
- Use grids, guides, and snapping features as design aids
- Adapt to Affinity Designer's custom workspaces and keyboard shortcuts
- Explore the workflow and design best practices for more predictable and successful outcomes
- Identify potential stumbling blocks in your design process and learn how to avoid them

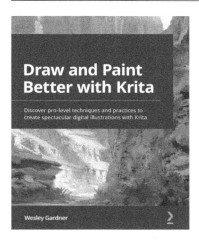

Draw and Paint Better with Krita

Wesley Gardner

ISBN: 978-1-80107-176-5

- Use layers, layer management, and layer blending modes to make images pop
- Understand Krita's default workspace and customize it
- Understand the terminology of digital visual communication (dots per inch, resolution, and more)
- Explore color in a digital space, such as RGB profiles and Look-Up-Tables (LUTS)
- Discover the color wheel for painting and learn how digital color (light and alpha channels) works as opposed to traditional painting materials
- Focus on proper layer management for easy, non-destructive manipulation of art pieces quickly

Packt is searching for authors like you

If you're interested in becoming an author for Packt, please visit `authors.packtpub.com` and apply today. We have worked with thousands of developers and tech professionals, just like you, to help them share their insight with the global tech community. You can make a general application, apply for a specific hot topic that we are recruiting an author for, or submit your own idea.

Share Your Thoughts

Now you've finished *Adobe Illustrator for Creative Professionals*, we'd love to hear your thoughts! Scan the QR code below to go straight to the Amazon review page for this book and share your feedback or leave a review on the site that you purchased it from.

`https://packt.link/r/1-800-56925-4`

Your review is important to us and the tech community and will help us make sure we're delivering excellent quality content.